An Account Of The Customs And Manners Of The Micmakis And Maricheets Savage Nations, Now Dependent On The Government Of Cape-Breton

By

Antoine Simon Maillard

Double9 BOOKS

An Account Of The Customs And Manners Of The Micmakis And Maricheets Savage Nations, Now Dependent On The Government Of Cape-Breton

by Antoine Simon Maillard

ISBN:978-93-59951-99-7

Published by

DOUBLE 9 BOOKS

2/13-B, Ansari Road
Daryaganj, New Delhi – 110002
info@double9books.com
www.double9books.com
Tel. 011-40042856

ABOUT THE AUTHOR

Abbé Pierre Antoine Simon Maillard (1710–12 August 1762) was a priest of French origin. He is best known for his efforts to the development of a writing system for the Mi'kmaq of Île Royale, New France (now Cape Breton Island, Canada). He is also recognized with assisting in the negotiation of a peace pact between the British and the Mi'kmaq, which resulted in the Burying of the Hatchet ceremony (Nova Scotia). He was the first Catholic priest in Halifax, Nova Scotia, and he is buried in St. Peter's Cemetery in the heart of the city. Maillard was born in 1710 in the diocese of Chartres, France. He acquired his theological education at Paris' Séminaire de Saint-Esprit. In 1734, the Abbé de L'Isle-Dieu chose Maillard from a group of seminarists loaned to the Séminaire des Missions Étrangères, which was in need of manpower. Maillard was chosen in 1735 for the Mi'kmaq missions on Île Royale (now Cape Breton Island) after spending eight months in that institution. According to his recommendation letter, "he is a young priest who has greatly edified us full of zeal and piety." On August 13, 1735, Maillard arrived in Fortress Louisbourg aboard the ship Rubis. He spent a lot of time with the Mi'kmaq people.

CONTENTS

PREFACE

For the better understanding of the letter immediately following, it may not be unnecessary to give the reader some previous idea of the people who are the subject of it, as well of the letter-writer.

The best account of the *Mickmakis* I could find, and certainly the most authentic, is in a memorial furnished by the French ministry in April, 1751, from which the following paragraph is a translated extract:

"The government of the savages dependent on Cape-Breton exacts a particular attention. All these savages go under the name of *Mickmakis*. Before the last war they could raise about six hundred fighting-men, according to an account given in to his most Christian majesty, and were distributed in several villages established on Cape-Breton island, island of St. John, on both the coasts of Acadia (Nova-Scotia) and on that of Canada. All, or most of the inhabitants of these villages have been instructed in the Christian religion, by missionaries which the king of France constantly maintains amongst them. It is customary to distribute every year to them presents, in the name of his majesty, which consist in arms, ammunition of war, victuals, cloathing, and utensils of various sorts. And these presents are regulated according to the circumstances of the time, and to the satisfaction that shall have been given to the government by the conduct of these savages. In the last war they behaved so as to deserve our approbation, and indeed have, on all occasions, given marks of their attachment and fidelity. Since the peace too, they have equally distinguished themselves in the disturbances that are on foot on the side of Acadia (Nova-Scotia)."

The last part of this foregoing paragraph needs no comment. Every one knows by what sort of service these savages merit the encouragement of the French government, and by what acts of perfidy and cruelty exercised on the English, they are to earn their reward.

The *Maricheets*, mentioned in the said letter form a distinct nation, chiefly settled at St. John's, and are often confounded with the *Abenaquis*, so as to pass for one nation with them, though there is certainly some distinction. They used, till lately, to be in a constant state of hostility with the Mickmakis. But, however, these nations may be at peace or variance with one another, in one point they agree, which is a thorough enmity to the English, cultivated, with great application by the missionaries, who add to the scandal of a conduct so contrary to their profession, the baseness of denying or evading the charge by the most pitiful equivocations. It is with the words peace, charity, and universal benevolence, for ever in their mouths, that these incendiaries, by instigations direct and indirect, inflame and excite the savages to commit the cruellest outrages of war, and the blackest acts of treachery. Poor Captain How! is well known to have paid with his life, infamously taken away by them, at a parley, the influence one of these missionaries (now a prisoner in the island of Jersey,) had over these misguided wretches, whose native innocence and simplicity are not proof against the corruption, and artful suggestions of those holy seducers.

It would not, perhaps, be impossible for the English, if they were to apply proper means, and especially lenient ones, to recover the affections of these people, which, for many reasons, cannot be entirely rooted in the French interest. That great state-engine of theirs, religion, by which they have so strong a hold on the weak and credulous savages, might not, however, be an invincible bar to our success, if it was duly counter-worked by the offer of a much more pure and rational one of our own, joined to such temporal advantages as would shew them their situation capable of being much meliorated, in every respect; and especially that of freedom, which they cannot but be sensible, is daily decreasing under the insidious encroachments and blandishments of the French, who never cares but to enslave, nor hug but to stifle, whose pretences, in short, to superior humanity and politeness, are not amongst their least arts of conquest.

As to the letter-writer, he is an abbot much respected in those parts, who has resided the greatest part of his life amongst the Mickmakis, and is perfectly acquainted with their language, in the composing of a Dictionary of which he has labored eighteen or twenty years; but I cannot learn that it is yet published, and probably for reasons of state, it never may. The letter, of which the translation is now given, exists only in a manuscript, having never been printed, being entirely written for the satisfaction of a friend's curiosity, in relation to the original manners and customs of the people of which it treats, and which, being those of savages in the primitive state of unpolished nature, may perhaps, to a philosophical enquirer, afford more amusement and instruction than those of the most refined societies. What man really is, appears at least plainer in the uncultivated savage, than in the civilized European.

The account of Acadia (Nova-Scotia) will, it is to be hoped, appear not uncurious; allowance being made for its being only in form of a letter.

A LETTER, &c

Micmaki-Country **, March 27, 1755.**

SIR,

I should long before now have satisfied you in those points of curiosity you expressed, concerning the savages amongst whom I have so long resided, if I could have found leisure for it. Literally true it is, that I have no spare time here, unless just in the evening, and that not always. This was my case too in Louisbourg; and I do not doubt but you will be surprised at learning, that I enjoy as little rest here as there.

Had you done me, Sir, the honor of passing with me but three days only, you would soon have seen what sort of a nation it is that I have to deal with. I am obliged to hold frequent and long parleys with them, and, at every occasion, to heap upon them the most fair and flattering promises. I must incessantly excite them to the practice of acts of religion, and labor to render them tractable, sociable, and loyal to the king (of France). But especially, I apply myself to make them live in good understanding with the French.

With all this, I affect a grave and serious air, that awes and imposes upon them. I even take care of observing measure and cadence in the delivery of my words, and to make choice of those expressions the properest to strike their attention, and to hinder what I say from falling to the ground. If I cannot boast that my harangues have all the fruit and success that I could wish, they are not however wholly without effect. As nothing inchants those people more than a style of metaphors and allegories, in which even their common conversation abounds, I adapt myself to their taste, and never please them better than when I give what I say this turn, speaking to them in their own language. I borrow the most lively images from those objects of nature, with which they are so well acquainted; and am rather more regular than even themselves, in the arrangement of my phrases. I affect, above all, to rhime as they do, especially at each member of a period. This contributes to give them so great an idea of me, that they imagine this gift of speaking is rather an inspiration, than an acquisition by study and meditation. In truth, I may venture to say, without presumption, that I talk the *Micmaki* language as fluently, and as elegantly, as the best of their women, who most excel in this point.

Another of my occupations is to engage and spur them on to the making a copious chace, when the hunting-season comes in, that their debts to the dealers with them may be paid, their wives and children cloathed, and their credit supported.

It is neither gaming nor debauchery that disable them from the payment of their debts, but their vanity, which is excessive, in the presents of peltry they make to other savages, who come either in quality of envoys from one country to another, or as friends or relations upon a visit to one another. Then it is, that a village is sure to exhaust itself in presents; it being a standing rule with them, on the arrival of such persons, to bring out every thing that they have acquired, during the winter and spring season, in order to give the best and most advantageous idea of themselves. Then it is chiefly they make feasts, which sometimes last several days; of the manner of which I should perhaps spare you the description, if the ceremony that attends them did not include the strongest attestation of the great stress they lay on hunting; the excelling wherein they commonly take for their text in their panegyrics on these occasions, and consequently enters, for a great deal, into the idea you are to conceive of the life and manners of the savages in these parts.

The first thing I am to observe to you is, that one of the greatest dainties, and with which they crown their entertainments, is the flesh of dogs. For it is not till the envoys, friends, or relations, are on the point of departure, that, on the eve of that day, they make a considerable slaughter of dogs, which they slea, draw, and, with no other dressing, put whole into the kettle; from whence they take them half boiled, and carve out into as many pieces as there are guests to eat of them, in the cabbin of him who gives the treat. But every one, before entering the cabbin, takes care to bring with him his *Oorakin*, or bowl, made of bark of birch-tree, either polygone shaped, or quite round; and this is practised at all their entertainments. These pieces of dogs flesh are accompanied with a small *Oorakin* full of the oil or fat of seal, or of elk's grease, if this feast is given at the melting-time of the snow. Every one has his own dish before him, in which he sops his flesh before he eats it. If the fat be hard, he cuts a small piece of it to every bit of flesh he puts into his mouth, which serves as bread with us. At the end of this fine regale, they drink as much of the oil as they can, and wipe their hands on their hair. Then come in the wives of the master and persons invited, who carry off their husbands plates, and retire together to a separate place, where they dispatch the remains.

After grace being said by the oldest of the company, who also never fails of pronouncing it before the meal, the master of the treat appears as if buried in a profound contemplation, without speaking a word, for a full quarter of an hour; after which, waking as it were out of a deep sleep, he orders in the *Calumets*, or *Indian* pipes, with tobacco. First he fills his own, lights it, and, after sucking in two or three whiffs, he presents it to the most considerable man in the company: after which, every one fills his pipe and smoaks.

The calumets lighted, and the tobacco burning with a clear fire, are scarce half smoked out, before the man of note before mentioned (for the greatest honors being paid him) gets up, places himself in the midst of the cabbin, and pronounces a speech of thanksgiving. He praises the master of the feast, who has so well regaled him and all the company. He compares him to a tree, whose large and strong roots afford nourishment to a number of small shrubs; or to a salutary medicinal herb, found accidentally by such as frequent the lakes in their canoes. Some I have heard, who, in their winter-feasts, compared him to the turpentine-tree, that never fails of yielding its sap and gummy distillation in all seasons: others to those temperate and mild days, which are sometimes seen in the midst of the severest winter. They employ a thousand similies of this sort, which I omit. After this introduction, they proceed to make honorable mention of the lineage from which the matter of the feast is descended.

"How great (will the oldest of them say) art thou, through thy great, great, great grand-father, whose memory is still recent, by tradition, amongst us, for the plentiful huntings he used to make! There was something of miraculous about him, when he assisted at the beating of the woods for elks, or other beasts of the fur. His dexterity at catching this game was not superior to our's; but there was some unaccountable secret he particularly possessed in his manner of seizing those creatures, by springing upon them, laying hold of their heads, and transfixing them at the same time with his hunting-spear, though thrice as strong and as nimble again as he was, and much more capable with their legs only, than we with our rackets [a sort of buskined shoes made purposely for the Indian travels over the snow], to make their way over mountains of snow: he would nevertheless follow them, dart them, without ever missing his aim, tire them out with his chace, bring them down, and mortally wound them. Then he would regale us with their blood, skin them, and deliver up the carcass to us to cut to pieces. But if thy great, great, great grand-father made such a figure in the chace, what has not thy great, great grand-father done with respect to the beavers, those animals almost men? whose industry he surpassed by his frequent watchings round their cabbins, by the repeated alarms he would give them several times in one evening, and oblige them thereby to return home, so that he might be sure of the number of those animals he had seen dispersed during the day, having a particular foresight of the spot to which they would come to load their tails with earth, cut down with their teeth such and such trees for the construction of their huts. He had a particular gift of knowing the favorite places of those animals for building them. But now let us rather speak of your great grand-father, who was so expert at making of snares for moose-deer, martins, and elks. He had particular secrets, absolutely unknown to any but himself, to compel these sort of creatures to run sooner into his snares than those of others; and he was accordingly always so well provided with furs, that he was never at a loss to oblige his friends. Now let us come to your grand-father, who has a thousand and a thousand times regaled the youth of his time with seals. How often in our young days have we greased our hair in his cabbin? How often have we been invited, and even compelled by his friendly violence, to go home with him, whenever we returned with our canoes empty, to be treated with seal, to drink the oil, and anoint ourselves with it? He even pushed his generosity so far, as to give us of the oil to take home with us. But now we are come to your father: there was a man for you! He used to signalize himself in every branch of chace; but especially in the art of shooting the game whether flying or sitting. He never missed his aim. He was particularly admirable for decoying of bustards by his artificial imitations. We are all of us tolerably expert at counterfeiting the cry of those birds; but as to him, he surpassed us in certain inflexions, of his voice, that made it impossible to distinguish his cry from that of the birds themselves. He had, besides, a particular way of motion with his body, that at a distance might be taken for the clapping of their wings, insomuch that he has often deceived ourselves, and put us to confusion, as he started out of his hiding-place.

"As for thyself, I say nothing, I am too full of the good things thou hast feasted me with, to treat on that subject; but I thank thee, and take thee by the hand, leaving to my fellow-guests the care of acquitting themselves of that duty."

After this, he sits down, and some other younger, and in course of less note, for they pay great respect to age, gets up, and makes a summary recapitulation of what the first speaker has said; commending his manner of singing the praises of the master of the feast's ancestors: to which he observes, there is nothing to be added; but that he has, however, left him one part of the task to be accomplished, which is, not to pass over in silence the feast to which he and the rest of his brethren are invited; neither to omit the merit and praises of him who has given the entertainment. Then quitting his place, and advancing in cadence, he takes the master of the treat by the hand, saying, "All the praises my tongue is about to utter, have thee for their object. All the steps I am going to take, as I dance lengthwise and breadthwise in thy cabbin, are to prove to thee the gaiety of my heart, and my gratitude. Courage! my friends, keep time with your motions and voice, to my song and dance."

With this he begins, and proceeds in his *Netchkawet*, that is, advancing with his body strait erect, in measured steps, with his arms a-kimbo. Then he delivers his words, singing and trembling with his whole body, looking before and on each side of him with a steady countenance, sometimes moving with a slow grave pace, then again with a quick and brisk one.

The syllables he articulates the most distinctly are, *Ywhannah, Owanna, Haywanna, yo! ha! yo! ha!* and when he makes a pause he looks full at the company, as much as to demand their chorus to the word *Heh!* which he pronounces with great emphasis. As he is singing and dancing they often repeat the word *Heh!* fetched up from the depth of their throat; and when he makes his pause, they cry aloud in chorus, *Hah!*

After this prelude, the person who had sung and danced recovers his breath and spirits a little, and begins his harangue in praise of the maker of the feast. He flatters him greatly, in attributing to him a thousand good qualities he never had, and appeals to all the company for the truth of what he says, who are sure not to contradict him, being in the same circumstance as himself of being treated, and answer him by the word *Heh*, which is as much as to say, *Yes*, or *Surely*. Then he takes them all by the hand, and begins his dance again: and sometimes this first dance is carried to a pitch of madness. At the end of it he kisses his hand, by way of salute to all the company; after which he goes quietly to his place again. Then another gets up to acquit himself of the same duty, and so do successively all the others in the cabbin, to the very last man inclusively.

This ceremony of thanksgiving being over by the men, the girls and women come in, with the oldest at the head of them, who carries in her left hand a great piece of birch-bark of the hardest, upon which she strikes as it were a drum; and to that dull sound which the bark returns, they all dance, spinning round on their heels, quivering, with one hand lifted, the other down: other notes they have none, but a guttural loud aspiration of the word Heh! Heh! Heh! as often as the old female savage strikes her bark-drum. As soon as she ceases striking, they set up a general cry, expressed by Yah! Then, if their dance is approved, they begin it again; and when weariness obliges the old woman to withdraw, she first pronounces her thanksgiving in the name of all the girls and women there. The introduction of which is too curious to omit, as it so strongly characterises the sentiments of the savages of that sex, and confirms the general observation, that where their bosom once harbours cruelty, they carry it greater lengths than even the men, whom frequently they instigate to it.

"You men! who look on me as of an infirm and weak sex, and consequently of all necessity subordinate to you, know that in what I am, the Creator has given to my share, talents and properties at least of as much worth as your's, I have had the faculty of bringing into the world warriors, great hunters, and admirable managers of canoes. This hand, withered as you see it now, whose veins represent the root of a tree, has more than once struck a knife into the hearts of the prisoners, who were given up to me for my sport. Let the river-sides, I say, for I call them to witness for me, as well as the woods of such a country, attest their having seen me more than once tear out the heart, entrails, and tongue, of those delivered up to me, without changing color, roast pieces of their flesh, yet palpitating and warm with life, and cram them down the throats of others, whom the like fate awaited. With how many scalps have not I seen my head adorned, as well as those of my daughters! With what pathetic exhortations have not I, upon occasion, rouzed up the spirit of our young men, to go in quest of the like trophies, that they might atchieve the reward, honor, and renown annexed to the acquisition of them: but it is not in these points alone that I have signalized myself. I have often brought about alliances, which there was no room to think could ever be made; and I have been so fortunate, that all the couples whose marriages I have procured, have been prolific, and furnished our nation with supports, defenders, and subjects, to eternize our race, and to protect us from the insults of our enemies. These old firs, these antient spruce-trees, full of knots from the top to the root, whose bark is falling off with age, and who yet preserve their gum and powers of life, do not amiss resemble me. I am no longer what I was; all my skin is wrinkled and furrowed, my bones are almost every where starting through it. As to my outward form, I may well be reckoned amongst the things, fit for nothing but to be totally neglected and thrown aside; but I have still within me wherewithal to attract the attention of those who know me."

After this introduction follow the thanksgiving and encomiums, much in the same taste as the first haranguer's amongst the guests. This is what is practised in all the more solemn entertainments, both on the men and women's side. Nor can you imagine, how great an influence such praises have over them, derived as they are from the merit of hunting, and how greatly they contribute to inflame their passion for it. Nor is it surprising, considering how much almost the whole of their livelihood depends upon the game of all sorts that is the object of their chace.

They have also a kind of feasts, which may be termed war-feasts, since they are never held but in time of war, declared, commenced, or resolved. The forms of these are far different from those of pacific and friendly entertainments. There is a mixture of devotion and ferocity in them, which at the same time that it surprises, proves that they consider war in a very solemn light, and as not to be begun without the greatest reason and justice; which motives, once established, or, which is the same thing, appearing to them established, there is nothing they do not think themselves permitted against their enemy, from whom they, on the other hand, expect no better quarter than they themselves give.

To give you an idea of their preparatory ceremony for a declaration of war, I shall here select for you a recent example, in the one that broke out not long ago between the Micmaquis, and Maricheets. These last had put a cruel affront on the former, the nature of which you will see in the course of the following description: but I shall call the Micmaquis the aggressors, because the first acts of hostility in the field began from them. Those who mean to begin the war, detach a certain number of men to make incursions on the territories of their enemies, to ravage the country, to destroy the game on it, and ruin all the beaver-huts they can find on their rivers and lakes, whether entirely, or only half-built. From this expedition they return laden with game and peltry; upon which the whole nation assembles to feast on the meat, in a manner that has more of the carnivorous brute in it than of the human creature. Whilst they are eating, or rather devouring, all of them, young and old, great and little, engage themselves by the sun, the moon, and the name of their ancestors, to do as much by the enemy-nation.

When they have taken care to bring off with them a live beast, from the quarter in which they have committed their ravage, they cut its throat, drink its blood, and even the boys with their teeth tear the heart and entrails to pieces, which they ravenously devour, giving thereby to understand, that those of the enemies who shall fall into their hands, have no better treatment to expect at them.

After this they bring out *Oorakins*, (bowls of bark) full of that coarse vermillion which is found along the coast of Chibucto, and on the west-side of Acadia (Nova-Scotia) which they moisten with the blood of the animal if any remains, and add water to compleat the dilution. Then the old, as well as the young, smear their faces, belly and back with this curious paint; after which they trim their hair shorter, some of one side of the head, some of the other; some leave only a small tuft on the crown of their head; others cut their hair entirely off on the left or right side of it; some again leave nothing on it but a lock, just on the top of their forehead, and of the breadth of it, that falls back on the nape of the neck. Some of them bore their ears, and pass through the holes thus made in them, the finest fibril-roots of the fir, which they call *Toobee*, and commonly use for thread; but on this occasion serve to string certain small shells. This military masquerade, which they use at once for terror and disguise, being compleated, all the peltry of the beasts killed in the enemy's country, is piled in a heap; the oldest *Sagamo*, or chieftain of the assembly gets up, and asks, "What weather it is? Is the sky clear? Does the sun shine?" On being answered in the affirmative, he orders the young men to carry the pile of peltry to a rising-ground, or eminence, at some little distance from the cabbin, or place of assembly. As this is instantly done, he follows them, and as he walks along begins, and continues his address to the sun in the following terms:

"Be witness, thou great and beautiful luminary, of what we are this day going to do in the face of thy orb! If thou didst disapprove us, thou wouldst, this moment, hide thyself, to avoid affording the light of thy rays to all the actions of this assembly. Thou didst exist of old, and still existeth. Thou remainest for ever as beautiful, as radiant, and as beneficent, as when our first fore fathers beheld thee. Thou wilt always be the same. The father of the day can never fail us, he who makes every thing vegetate, and without whom cold, darkness, and horror, would every where prevail. Thou knowest all the iniquitous procedure of our enemies towards us. What perfidy have they not used, what deceit have they not employed, whilst we had no room to distrust them? There are now more than five, six, seven, eight moons revolved since we left the principal amongst our daughters with them, in order thereby to form the most durable alliance with them, (for, in short, we and they are the same thing as to our being, constitution, and blood); and yet we have seen them look on these girls of the most distinguished rank, *Kayheepidetchque*, as mere playthings for them, an amusement, a pastime put by us into their hands, to afford them a quick and easy consolation, for the fatal blows we had given them in the preceding war. Yet, we had made them sensible, that this supply of our principal maidens was, in order that they should re-people their country more honorably, and to put them under a necessity of conviction, that we were now become sincerely their friends, by delivering to them so sacred a pledge of amity, as our principal blood. Can we then, unmoved, behold them so basely abusing that thorough confidence of ours? Beautiful, all-seeing, all-penetrating luminary! without whose influence the mind of man has neither efficacy nor vigor, thou hast seen to what a pitch that nation (who are however our brothers) has carried its insolence towards our principal maidens. Our resentment would not have been so extreme with respect to girls of more common birth, and the rank of whose fathers had not a right to make such an impression on us. But here we are wounded in a point there is no passing over in silence or unrevenged. Beautiful luminary! who art thyself so regular in thy course, and in the wise distribution thou makest of thy light from morning to evening, wouldst thou have us not imitate thee? And whom can we better imitate? The earth stands in need of thy governing thyself as thou dost towards it. There are certain places, where thy influence does not suffer itself to be felt, because thou dost not judge them worthy of it. But, as for us, it is plain that we are thy children; for we can know no origin but that which thy rays have given us, when first marrying efficaciously, with the earth we inhabit, they impregnated its womb, and caused us to grow out of it like the herbs of the field, and the trees of the forest, of which thou art equally the common father. To imitate thee then, we cannot do better than no longer to countenance or cherish those, who have proved themselves so unworthy thereof. They are no longer, as to us, under a favorable aspect. They shall dearly pay for the wrong they have done us. They have not, it is true, deprived us of the means of hunting for our maintenance and cloathing; they have not cut off the free passage of our canoes, on the lakes and rivers of this country; but they have done worse; they have supposed in us a tameness of sentiments, which does not, nor cannot, exist in us. They have defloured our principal maidens in wantonness, and lightly sent them back to us. This is the just motive which cries out for our vengeance. Sun! be thou favorable to us in this point, as thou art in that of our hunting, when we beseech thee to guide us in quest of our daily support. Be propitious to us, that we may not fail of discovering the ambushes that may be laid for us; that we may not be surprized unawares in our cabbins, or elsewhere; and, finally, that we may not fall into the hands of our enemies. Grant them no chance with us, for they deserve none. Behold the skins of their beasts now a burnt-offering to thee! Accept it, as if the fire-brand I hold in my hands, and now set to the pile, was lighted immediately by thy rays, instead of our domestic fire."

Every one of the assistants, as well men as women, listen attentively to this invocation, with a kind of religious terror, and in a profound silence. But scarce is the pile on a blaze, but the shouts and war-cries begin from all parts. Curses and imprecations are poured forth without mercy or reserve, on the enemy-nation. Every one, that he may succeed in destroying any particular enemy he may have in the nation against which war is declared, vows so many skins or furs to be burnt in the same place in honor of the sun. Then they bring and throw into the fire, the hardest stones they can find of all sizes, which are calcined in it. They take out the properest pieces for their purpose, to be fastened to the end of a stick, made much in the form of a hatchet-handle. They slit it at one end, and fix in the cleft any fragment of those burnt stones, that will best fit it, which they further secure, by binding it tightly round with the strongest *Toobee*, or fibrils of fir-root above-mentioned; and then make use of it, as of a hatchet, not so much for cutting of wood, as for splitting the skull of the enemy, when they can surprize him. They form also other instruments of war; such as long poles, one of which is armed with bone of elk, made pointed like a small-sword, and edge of both sides, in order to reach the enemy at a distance, when he is obliged to take to the woods. The arrows are made at the same time, pointed at the end with a sharp bone. The wood of which these arrows are made, as well as the bows, must have been dried at the mysterious fire, and even the guts of which the strings are made. But you are here to observe, I am speaking of an incident that happened some years ago; for, generally speaking, they are now better provided with arms, and iron, by the Europeans supplying them, for their chace, in favor of their dealings with them for their peltry. But to return to my narration.

Whilst the fire is still burning, the women come like so many furies, with more than bacchanalian madness, making the most hideous howlings, and dancing without any order, round the fire. Then all their apparent rage turns of a sudden against the men. They threaten them, that if they do not supply them with scalps, they will hold them very cheap, and look on them as greatly inferior to themselves; that they will deny themselves to their most lawful pleasures; that their daughters shall be given to none but such as have signalized themselves by some military feat; that, in short, they will themselves find means to be revenged of them, which cannot but be easy to do on cowards.

The men, at this, begin to parley with one another, and order the women to withdraw, telling them, that they shall be satisfied; and that, in a little time, they may expect to have prisoners brought to them, to do what they will with them.

The next thing they agree on is to send a couple of messengers, in the nature of heralds at arms, with their hatchets, quivers, bows, and arrows, to declare war against the nation by whom they conceive themselves aggrieved. These go directly to the village where the bulk of the nation resides, observing a sullen silence by the way, without speaking to any that may meet them. When they draw near the village, they give the earth several strokes with their hatchets, as a signal of commencing hostilities in form; and to confirm it the more, they shoot two of their best arrows at the village, and retire with the utmost expedition. The war is now kindled in good earnest, and it behoves each party to stand well on its guard. The heralds, after this, return to make a report of what they have done; and to prove their having been at the place appointed, they do not fail of bringing away with them some particular marks of that spot of the country. Then it is, that the inhabitants of each nation begin to think seriously, whether they shall maintain their ground by staying in their village, and fortifying it in their manner, or look out for a place of greater safety, or go directly in quest of the enemy. Upon these questions they assemble, deliberate, and hold endless consultations, though withal not uncurious ones: for it is on these occasions, that those of the greatest sagacity and eloquence display all their talents, and make themselves distinguished. One of their most common stratagems, when there were reasons for not attacking one another, or coming to a battle directly, was for one side to make as if they had renounced all thoughts of acting offensively. A party of those who made this feint of renunciation, would disperse itself in a wood, observing to keep near the borders of it; when, if any stragglers of the enemy's appeared, some one would counterfeit to the life the particular cry of that animal, in the imitation of which he most excelled; and this childish decoy would, however, often succeed, in drawing in the young men of the opposite party into their ambushes.

Sometimes the scheme was to examine what particular spot lay so, that the enemies must, in all necessity, pass through it, to hunt, or provide bark for making their canoes. It was commonly in these passes, or defiles, that the bloodiest encounters or engagements happened, when whole nations have been known to destroy one another, with such an exterminating rage on both sides, that few have been left alive on either; and to say the truth, they were, generally speaking, mere cannibals. It was rarely the case that they did not devour some limbs, at least, of the prisoners they made upon one another, after torturing them to death in the most cruel and shocking manner: but they never failed of drinking their blood like water; it is now, some time, that our Micmakis especially are no longer in the taste of exercising such acts of barbarity. I have, yet, lately myself seen amongst them some remains of that spirit of ferocity; some tendencies and approaches to those inhumanities; but they are nothing in comparison to what they used to be, and seem every day wearing out. The religion to which we have brought them over, and our remonstrances have greatly contributed to soften that savage temper, and atrocious vindictiveness that heretofore reigned amongst them. But remember, Sir, that as to this point I am now only speaking, upon my own knowlege, of the Micmakis and Mariquects, who, though different in language, have the same customs and manners, and are of the same way of thinking and acting.

But to arrive at any tolerable degree of conjecture, whence these people derive their origin own myself at a loss: possibly some light might be got into it, by discovering whether there was any affinity or not between their language, and that of the Orientalists, as the Chinese or Tartars. In the mean time, the abundance of words in this language surprized, and continues to surprize me every day the deeper I get into it. Every thing is proper in it; nothing borrowed, as amongst us. Here are no auxiliary verbs. The prepositions are in great number. This it is that gives great ease, fluency, and richness to the expression of whatever you require, when you are once master enough to join them to the verbs. In all their absolute verbs they have a dual number. What we call the imperfect, perfect, and preter-perfect tenses of the indicative mood, admits, as with us, of varied inflexions of the terminations to distinguish the person; but the difference of the three tenses is express, for the preter-perfect by the preposition *Keetch*; for the preter-pluperfect by *Keetch Keeweeh*: the imperfect is again distinguished from them by having no preposition at all.

They have no feminine termination, either for the verbs or nouns. This greatly facilitates to me my composition of songs and hymns for them, especially as their prose itself naturally runs into poetry, from the frequency of their tropes and metaphors; and into rhime, from their nouns being susceptible of the same termination, as that of the words in the verbs which express the different persons. In speaking of persons absent, the words change their termination, as well in the nouns as in the verbs.

They have two distinctions of style; the one noble, or elevated, for grave and important subjects, the other ignoble, or trivial, for familiar or vulgar ones. But this distinction is not so much with them, as with us, marked by a difference of words, but of terminations. Thus, when they are treating of solemn, or weighty matters, they terminate the verb and the noun by another inflexion, than what is used for trivial or common conversation.

I do not know, whether I explain clearly enough to you this so material a point of their elocution; but it makes itself clearly distinguished, when once one comes to understand the language, in which it supplies the place of the most pathetic emphasis, though even that they do not want, nor great expression in their gestures and looks. All their conjugations are regular and distinct.

Yet, with all these advantages of language, the nation itself is extreamly ignorant as to what concerns itself, or its origin, and their traditions are very confused and defective. They know nothing of the first peopling of their country, of which they imagine themselves the Aborigines. They often talk of their ancestors, but have nothing to say of them that is not vague or general. According to them, they were all great hunters, great wood-rangers, expert managers of canoes, intrepid warriors, that took to wives as many as they could maintain by hunting. They had too a custom amongst them, that if a woman grew pregnant whilst she was sucking a child, they obliged her to use means for procuring an abortion, in favor of the first-come, who they supposed would otherwise be defrauded of his due nourishment. Most of them also value themselves on being descended from their Jugglers, who are a sort of men that pretend to foretel futurity by a thousand ridiculous contorsions and grimaces, and by frightful and long-winded howlings.

The great secret of these Jugglers consists in having a great *Oorakin* full of water, from any river in which it was known there were beaver-huts. Then he takes a certain number of circular turns round this Oorakin, as it stands on the ground, pronouncing all the time with a low voice, a kind of gibberish of broken words, unintelligible to the assistants, and most probably so to himself, but which those, on whom he means to impose, believe very efficacious. After this he draws near to the bowl, and bending very low, or rather lying over it, looks at himself in it as in a glass. If he sees the water in the least muddy, or unsettled, he recovers his erect posture, and begins his rounds again, till he finds the water as clear as he could wish it for his purpose, and then he pronounces over it his magic words. If after having repeated them twice or thrice, he does not find the question proposed to him resolved by this inspection of the water, nor the wonders he wants operated by it, he says with a loud voice and a grave tone, that the *Manitoo*, or *Miewndoo*, (the great spirit) or genius, which, according to them, has all knowledge of future events, would not declare himself till every one of the assistants should have told him (the Juggler) in the ear what were his actual thoughts, or greatest secret. [A Romish missionary must, with a very bad grace, blame the Jugglers, for what himself makes such a point of religion in his *auricular confession*. Even the appellation of *Juggler* is not amiss applicable to those of their craft, considering all their tricks and mummery not a whit superior to those of these poor savages, in the eyes of common-sense. Who does not know, that the low-burlesque word of *Hocus-pocus*, is an humorous corruption of their *Hoc est corpus meum*, by virtue of which, they make a *God* out of a vile wafer, and think it finely solved, by calling it a *mystery*, which, by the way is but another name for *nonsense*. Is there any thing amongst the savages half so absurd or so impious?] To this purpose he gets up, laments, and bitterly inveighs against the bad dispositions of those of the assistants, whose fault it was, that the effects of his art were obstructed. Then going round the company, he obliges them to whisper him in the ear, whatever held the first place in their minds; and the simplicity of the greater number is such, as to make them reveal to him what it would be more prudent to conceal. By these means it is, that these artful Jugglers renders themselves formidable to the common people, and by getting into the secrets of most of the families of the nation, acquire a hank over them. Some, indeed, of the most sensible see through this pitiful artifice, and look on the Jugglers in their proper light of cheats, quacks, and tyrants; but out of fear of their established influence over the bulk of the nation, they dare not oppose its swallowing their impostures, or its regarding all their miserable answers as so many oracles. When the Juggler in exercise, has collected all that he can draw from the inmost recesses of the minds of the assistants, he replaces himself, as before, over the mysterious bowl of water, and now knows what he has to say. Then, after twice or thrice laying his face close to the surface of the water, and having as often made his evocations in uncouth, unintelligible words, he turns his face to his audience, sometimes he will say, "I can only give a half-answer upon such an article; there is an obstacle yet unremoved in the way, before I can obtain an entire solution, and that is, there are some present here who are in such and such a case. That I may succeed in what is asked of me, and that interests the whole nation, I appoint that person, without my knowing, as yet, who it is, to meet me at such an hour of the night. I name no place of assignation but will let him know by a signal of lighted fire, where he may come to me, and suffer himself to be conducted wherever I shall carry him. The *Manitoo* orders me to spare his reputation, and not expose him; for if there is any harm in it to him, there is also harm to me."

Thus it is the Juggler has the art of imposing on these simple credulous creatures, and even often succeeds by it in his divinations. Sometimes he does not need all this ceremonial. He pretends to foretell off-hand, and actually does so, when he is already prepared by his knowledge, cunning, or natural penetration. His divinations chiefly turn on the expedience of peace with one nation, or of war with another; upon matches between families, upon the long life of some, or the short life of others; how such and such persons came by their deaths, violently or naturally; whether the wife of some great *Sagamo* has been true to his bed or not; who it could be that killed any particular persons found dead of their wounds in the woods, or on the coast. Sometimes they pretend it's the deed of the *Manitoo*, for reasons to them unknown: this last incident strikes the people with a religious awe. But what the Jugglers are chiefly consulted upon, and what gives them the greatest credit, is to know whether the chace of such a particular species of beasts should be undertaken; at what season, or on which side of the country; how best may be discovered the designs of any nation with which they are at war; or at what time such or such persons shall return from their journey. The Juggler pretends to see all this, and more, in his bowl of water: divination by coffee-grounds is a trifle to it. He is also applied to, to know whether a sick person shall recover or die of his illness. But what I have here told you of the procedure of these Jugglers, you are to understand only of the times that preceded the introduction of Christianity amongst these people, or of those parts where it is not yet received: for these practices are no longer suffered where we have any influence.

Amongst the old savages lately baptized, I could never, from the accounts they gave me of the belief of their ancestors, find any true *knowledge* of the supreme Being; no idea, I mean, approaching to that we have, or rather nothing but a vague imagination. They have, it is true, a confused notion of a Being, acting they know not how [Who does?], in the universe, but they do not make of him a great soul diffused through all its parts. They have no conception or knowledge of all the attributes we bestow on the Deity. Whenever they happen to philosophize upon this *Manitoo*, or great spirit, they utter nothing but *rêveries* and absurdities. [Are not there innumerable volumes on this subject, to which the same objection might as justly be made? Possibly the savages, and the deepest divines, with respect to the manner of the Deity's existence, may have, in point of ignorance, nothing to reproach one another. It matters very little, whether one sees the sun from the lowest valley, or the highest mountain, when the immensity of its distance contracts the highest advantage of the eminence to little less than nothing. Surely the infinite superiority of the Deity, must still more effectually mock the distinction of the mental eye, at the same time that his existence itself is as plain as that of the sun, and like that too, dazzling those most, who contemplate it most fixedly; reduces them to close the eye, not to exclude the light, but as overpowered by it.]

Amongst other superstitious notions, not the least prevalent is that of the *Manitoo*'s exercise of his power over the dead, whom he orders to appear to them, and acquaint them with what passes at a distance, in respect to their most important concerns; to advise them what they had best do, or not do; to forewarn them of dangers, or to inspire them with revenge against any nation that may have insulted them, and so forth.

They have no idea of his spirituality, or even of the spirituality of that principle, which constitutes their own vital principle. They have even no word in their language that answers to that of soul in ours. The term approaching nearest thereto that we can find, is *M'cheejacmih,* which signifies *Shade,* and may be construed something in the nature of the *Manes* of the Romans.

The general belief amongst them is, that, after death, they go to a place of joy and plenty, in which sensuality is no more omitted than in Mahomet's paradise. There they are to find women in abundance, a country thick of all manner of game to humor their passion for hunting, and bows and arrows of the best sort, ready made. But these regions are supposed at a great distance from their's, to which they will have to travel; and therefore it's requisite to be well-provided, before they quit their own country, with arrows, long poles fit for hunting, or for covering cabbins, with bear-skins, or elk-hides, with women, and with some of their children, to make their journey to that place more commodious, more pleasant, and appear more expeditious. It was especially in character for a warrior, not to leave this world without taking with him some marks of his bravery, as particularly scalps. Therefore it was, that when any of them died, he was always followed by, at least, one of his children, some women, and above all, by her whom in his life he had most loved, who threw themselves into the grave, and were interred with him. They also put into it great strips, or rolls of the bark of birch, arrows, and scalps. Nor do they unfrequently, at this day, light upon some of these old burying-places in the woods, with all these funeral accompanyments; but of late, the interment of live persons has been almost entirely disused.

I never could learn whether they had any set formulary of prayer, or invocation to the *great Manitoo*; or whether they made any sacrifices of beasts or peltry, to any other *Manitoo*, in contradiction to him, or to any being whom they dreaded as an evil genius. I could discover no more than what I have above related of the ceremonies in honor of the sun. I know, indeed, they have a great veneration for the moon, which they invoke, whenever, under favor of its light, they undertake any journeys, either by land or water, or tend the snares they have set for their game. This is the prayer they occasionally address to it:

"How great, O moon! is thy goodness, in actually, for our benefit, supplying the place of the father of the day, as, next to him, thou hast concurred to make us spring out of that earth we have inhabited from the first ages of the world, and takest particular care of us, that the malignant air of the night, should not kill the principle and bud of life within us. Thou regardest us, in truth, as thy children. Thou hast not, from the first time, discontinued to treat us like a true mother. Thou guidest us in our nocturnal journies. By the favor of thy light it is, that we have often struck great strokes in war; and more than once have our enemies had cause to repent their being off their guard in thy clear winter-nights. Thy pale rays have often sufficiently lighted us, for our marching in a body without mistaking our way; and have enabled us not only to discover the ambushes of the enemy, but often to surprize him asleep. However we might be wanting to ourselves, thy regular course was never wanting to us. Beautiful spouse of the sun! give us to discover the tracks of elks, moose-deer, martins, lynxes, and bears, when urged by our wants, we pursue by night the hunt after these beasts. Give to our women the strength to support the pains of child-birth [*Lucina fer opem,* was also the cry amongst the ancient heathens], render their wombs prolific, and their breasts inexhaustible fountains."

I have often tried to find out, whether there was any tradition or knowledge amongst them of the deluge, but always met with such unsatisfactory answers, as entirely discouraged my curiosity on that head.

This nation counts its years by the winters. When they ask a man how old he is, they say, "How many winters have gone over thy head?"

Their months are lunar, and they calculate their time by them. When we would say, "I shall be six weeks on my journey;" they express it by, "I shall be a moon and a half on it."

Before *we* knew them, it was common to see amongst them, persons of both sexes of a hundred and forty, or a hundred and fifty years of age. But these examples of longevity are grown much more rare.

By all accounts too, their populousness is greatly decreased. Some imagine this is owing to that inveterate animosity, with which these so many petty nations were continually laboring one another's destruction and extirpation. Others impute it to the introduction by the Europeans, of the vice of drunkenness, and to the known effect of spirituous liquors in the excesses of their use, to which they are but too prone, in striking at the powers of generation, as well as at the principles of health and life. Not improbably too, numbers impatient of the encroachments of the Europeans on their country, and dreading the consequences of them to their liberty, for which they have a passionate attachment, and incapable of reconciling or assimilating their customs and manners to ours, have chosen to withdraw further into the western recesses of the continent, at a distance impenetrable to our approach.

But which ever of these conjectures is the truest, or whether or not all of these causes have respectively concurred, in a lesser or greater degree, the fact is certain, that all these northern countries are considerably thinned of their natives, since the first discovery of them by the Europeans. Nor have I reason to think, but that this is true of America in general, wherever they have carried their power, or extended their influence.

It is also true, that the women of this country are naturally not so prolific as those of some other parts of the world in the same latitude. One reason for this may be, their not having their menstrual flux so copiously, or for so long a time as those of Europe. Yet one would think, the plurality of wives permitted amongst them, might in some measure compensate for this defect, which, however, it evidently does not.

Their women have always observed, not to present themselves at any public ceremony, or solemnity, whilst under their monthly terms, nor to admit the embraces of their husbands.

At stated times they repair to particular places in the woods, where they recite certain formularies of invocation to the *Manitoo* dictated to them by some of their oldest *Sagamees*, or principal women, and more frequently by some celebrated Juggler of the village, that they may obtain the blessing of fruitfulness. For it is with them, as amongst the Jews, that barrenness is accounted opprobrious. A woman is not looked upon as a woman, till she has proved it, by her fulfilling what they consider as one of the great ends of her creation. Failing in that, she is divorced from her husband, and may then prostitute herself without any scandal. If she has no inclination or relish for this way of life, they compel her to it, in regard to their young men, who do not care to marry, till they are arrived at full-ripe years, and for whom, on their return from their warlike or hunting expeditions, they think it necessary to provide such objects of amusement. They pretend withal, that they are subject to insupportable pains in their loins, if such a remedy is not at hand to relieve them. But once more you are to remember, that I am only speaking of those people not yet converted to Christianity, by which this licentiousness is not allowed. And yet, notwithstanding the maxims we inculcate to them, the natives continue no other than what they were before, that is to say, as much addicted to venery as ever, and rarely miss an occasion of gratifying their appetite to it. The only way we can think of to prevent their offending religion, is to have them married as soon as they begin to feel themselves men. The restraint however in this point is, what they can least endure.

In their unconverted state, their manner of courtship and marriage is as follows: When a youth has an inclination to enter into the connubial state, his father, or next relation, looks out for a girl, to whose father the proposal is made: this being always transacted between the parents of the parties to be married. The young man, who is commonly about thirty years of age, or twenty at the least, rarely consults his own fancy in this point. The girl, who is always extreamly young, is never supposed to trouble her head about the measures that are taking to marry her. When the parents on each side have settled the matter, the youth is applied to, that he may prepare his calumet as soon as he pleases.

The calumet used on these occasions, is a sort of spungeous reed, which may furnish, according to its length, a number of calumets, each of which is about a foot long, to be lighted at one end, the other serving to suck in the smoak at the mouth, and is suffered to burn within an inch of the lips.

The speech made to the youth on this occasion is as follows: "Thou may'st go when thou wilt, by day or by night, to light thy calumet in such a cabbin. Thou must observe to direct the smoak of it towards the person who is designed for thee, and carry it so, that she may take such a taste to this vapor, as to desire of thee that she may smoak of thy calumet. Show thyself worthy of thy nation, and do honor to thy sex and youth. Suffer none in the cabbin to which thou art admitted, to want any thing thy industry, thy art, or thy arrows can procure them, as well for food, as for peltry, or oil, for the good of their bodies, inside and outside. Thou hast four winters given thee, for a trial of thy patience and constancy."

At this the youth never fails of going to the place appointed. If the girl, (who knows the meaning of this) has no particular aversion to him, she is soon disposed to ask his calumet of him. In some parts, but not in this where I am, she signifies her acceptance by blowing it out. Here she takes it from him, and sucking it, blows the smoak towards his nostrils, even sometimes so violently, as to make him qualm-sick, at which she is highly delighted. Nothing, however, passes farther against the laws of modesty, though she will tress his hair, paint his face, and imprint on various parts of his body curious devices and flourishes, all relative to their love; which she pricks in, and rubs over with a composition that renders the impression uncancellable.

If the parents of the girl are pleased with the procedure of the suitor, they commonly, at the end of the second year, dispense, in his favor, with the rest of the probation-time; and, indeed, they could not well before, the girl almost always wanting, from the time she is first courted, at least two years to bring on the age of consummation. They tell him, "Thou may'st now take a small part of the covering of thy beloved whilst she sleeps." No sooner is this compliment made him, than, without saying any thing, he goes out of the cabbin, armed with his bow and arrows, and hurrying home acquaints his friends, that he is going to the woods, whence he shall not return till it pleases his beloved to recall him.

Accordingly he repairs forthwith to the woods, and stays there for two or three days, diverting himself with hunting; at the end of which it has been agreed on, to send all the youths of the village to fetch him: and they come back loaded with game of all sorts, though the bridegroom is not suffered to carry any thing. There is also great provision made of seal and sea-cows for the wedding-feast.

The head Juggler of the village, meets the bridegroom who is at the head of the procession, takes him by the hand, and conducts him to the cabbin of the bride, where he is to take part of her bed; upon which he lies down by her side, and both continue unmoveable and silent like two statues, whilst they are obliged to hear the long tedious harangues of the Juggler, of the parents of both, and of their oldest relations. After that, they both get up, and are led, the one by the young men, the other by the girls, to the place of entertainment, all singing, shooting, and dancing.

The bridegroom is seated amongst the young men on one side, and the bride amongst the girls on another. One of his friends takes an *Oorakin*, loads it with roast-meat, and sets it down by him, whilst one of her's does the same thing, with an *Oorakin* of the same size, and nearly alike, which is placed by the bride's side. After this ceremony of placing the *Oorakin*, the Juggler pronounces certain magical words over the meat: he foretels, especially to the bride, the dreadful consequences she must expect from the victuals she is about to eat, if she has in her heart any perfidiousness towards her husband: that she may be assured of finding in the *Oorakin* that contains them, a certain prognostic of her future happiness, or unhappiness: of happiness, if she is disposed never in her life to betray her nation, nor especially her husband, upon any occasion, or whatever may befal her: of unhappiness, if through the caresses of strangers, or by any means whatever she should be induced to break her faith to him, or to reveal to the enemy the secrets of the country.

At the end of every period, all the assistants signify their assent to the Juggler's words, by a loud exclamation of *Hah!* Whilst he is talking, the particular friend of the bridegroom, and that of the bride, keep their eyes fixed on the two *Oorakins*; and as soon as he has done, the bride's friend making as if she did not think of what she was about, takes the *Oorakin* allotted for the bridegroom, and carries it to the bride, whilst the bridegroom's friend, (the thing being pre-concerted) acts the like mummery of inadvertence, and sets before the bridegroom the *Oorakin* belonging to the bride; after which the dishes are served in to the rest of the company. When they are all served, the two friends of the parties musing a little, pretend to have just then discovered their exchange of the bride and bridegroom's *Oorakins*. They declare it openly to each other, at which the Juggler takes up his cue, and with a solemn face says, "The *Manitoo* has had his designs in this mistake: he has vouchsafed to give an indubitable sign of his approbation of the strait alliance this day contracted. What is the one's, is the same as the other's. They are henceforward united, and are as one and the same person. It is done. May they multiply without end!" At this the assistants all start up, and with cries of joy, and congratulation, rush to embrace the bride and bridegroom, and overwhelm them with caresses. After which they sit very gravely down again to the entertainment before them, and dispatch it in great silence. This is followed by dances of all kinds, with which the feast for the day concludes, as must this letter, in which I have certainly had less attention to the observing the limits of one, than to the gratifying your curiosity, with respect to these people, amongst whom my lot has so long been cast.

I am, Sir,

Your most obedient

Humble servant,

To understand the following piece, it is necessary to know, that after the insidious peace of Aix-la-Chapelle, the savage nations, especially the Mickmakis and Maricheets continued hostilities against the English, at the underhand instigation of the French, who meant thereby to prevent, or at least distress, as much as obstruct, our new settlements in Nova-Scotia. For this purpose, the French missionaries had their cue from their government to act the incendiaries, and, to inflame matters to the highest pitch. These being, however, sensible, that the part assigned them was a very odious one, and inconsistent with the spirit of that religion for which they profess such zeal, one of them, by way of palliation, and in order to throw the blame on the English themselves, drew up the following state of the case, between our nation and the savages, viz.

MEMORIAL OF THE Motives of the Savages, called *Mickmakis* and *Maricheets*, for continuing the War with *England* since the last Peace

Dated *Isle-Royal*, 175-.

These nations have never been able to forget all that the English settled in North-America have done since the very first of their establishment, towards destroying them root and branch. They have especially, at every moment, before their eyes the following transactions:

In 1744, towards the end of October, Mr. Gorrhon, (perhaps Goreham) deceased, commanding a detachment of the English troops, sent to observe the retreat the French and savages were making from before Port-Royal (Annapolis) in Acadia, (Nova-Scotia): this detachment having found two huts of the Mickmaki-savages, in a remote corner, in which there were five women and three children, (two of the women were big with child) ransacked, pillaged, and burnt the two huts, and massacred the five women and three children. It is to be observed, that the two pregnant women were found with their bellies ripped open. An action which these savages cannot forget, especially as at that time they made fair war with the English. They have always looked on this deed as a singular mark of the most unheard-of cruelty. [Who would not look on it in the same light? But as no nation on earth is known to have more than ours constitutionally, a horror for such barbarities, especially in cold blood; it may be very easily presumed, that this fact was, if true, committed by some of the savages themselves, without the knowledge of the commander, or of any of the English troops.]

Five months before this action, one named *Danas*, or *David*, an English privateer, having treacherously hoisted French colors in the Streights of Fronsac, by means of a French deserter he had with him, decoyed on board his vessel the chief of the savages of Cape-Breton, called James Padanuque, with his whole family, whom he carried to Boston, where he was clapped into a dungeon the instant he was landed; from which he was only taken out to stifle him on board of a vessel, in which they pretended to return him safe to Cape-Breton. His son, at that time a boy of eight years of age, they will absolutely not release; though, since their detention of that young savage, they have frequently had prisoners sent back to them, without ransom, on condition of restoring the young man to his country: but though they accepted the condition, they never complied with it.

In the month of July, 1745, the same Danas, with the same success, employed the same decoy on a savage-family, which could not get out of their hands, but by escaping one night from their prisons.

About the same time one named Bartholomew Petitpas, an appointed savage-linguist, was carried away prisoner to Boston. The savages have several times demanded him in exchange for English prisoners they then had in their hands, of whom two were officers, to whom they gave their liberty, on condition of the Bostoners returning of Petitpas; whom, however, they not only kept prisoner, but afterwards put to death.

In the same year, 1745, a missionary of the savages of Cape Breton, Natkikouesch, Picktook, and of the island of St. John, having been invited by several letters, on the part of the commodore of the *English* squadron, and of the general of the land-forces, to a parley, those gentlemen desired with him, concerning the savages, repaired to Louisbourg, at that time in possession of the English, on the assurances they had given him in writing, and on the formal promises they had bound with an oath, of full liberty to return from whence he came, after having satisfied them in all they wanted of him. They detained him at Louisbourg, where they gave him a great deal of ill usage, and obliged him to embark, all sick as he was, and destitute of necessaries, on board of one of the ships of the squadron, in which he was conveyed to England, from whence he at length got to France. [Most probably he had not given the satisfaction required by those gentlemen, which had been confessedly by himself made the condition of his return.]

The same year, 1745, several bodies of the savages, deceased, and buried at *Port Tholouze,* were dug up again by the Bostoners, and thrown into the fire. The burying-place of the savages was demolished, and all the crosses, planted on the graves, broke into a thousand pieces.

In 1746, some stuffs that the savages had bought of the English, who then traded in the bay of Megagouetch at *Beau-bassin,* there being at that time a great scarcity of goods over all the country, were found to be *poisoned,* [Is it possible a missionary of the truths of the Gospel could gravely commit to paper such an infernal lie? If even the savages had been stupid enough of themselves to imbibe such a notion, was it not the duty of a Christian to have shewn them the folly of it, or even but in justice to the Europeans? But what must be their guilt, if they suggested it? Surely, scarce less than that of the action itself.] so that more than two hundred savages of both sexes perished thereby.

In 1749, towards the end of the month of May, at a time that the suspension of arms between the two crowns was not yet known in New France, the savages, having made prisoners two Englishmen of Newfoundland, had from these same prisoners the first news of the cessation of hostilities. They believed them on their bare words, expressed their satisfaction to them, treated them like brothers, unbound them, and carried them to their huts. The said prisoners rose in the night, and massacred twenty-five of these savages, men, women, and children. There were but two of the savages escaped this carnage, by being accidentally not present. [*How improbable is the whole of this story?*]

Towards the end of the same year, the English being come to Chibuckto, made the report be every where spread [The missionaries in those parts might indeed raise such reports; the which giving the savages an aversion to the English, forced them to take hostile measures against them in their own defence: but who would suspect the English themselves of raising them, in direct opposition to their own interest?], that they were going to destroy all the savages. They seemed to act in consequence thereto, since they sent detachments of their troops, on all sides, in pursuit of the savages.

These people were so alarmed with this procedure of the English, that from that time they determined, as weak as they were, to declare open war against them. Knowing that France had concluded a peace with England, they nevertheless resolved not to cease from falling on the English, wherever they could find them; saying, they were indispensably obliged to it, since, against all justice, they wanted to expel them out of their country. They then sent a declaration of war in form to the English, in the name of their nation, and of the savages in alliance with it.

As to what concerns the missionaries to the savages, they cannot be suspected of using any connivence in all this, if justice is done to the conduct they have always observed amongst them, and especially in the time of the last war. How many acts of inhumanity would have been committed by this nation, naturally vindictive, if the missionaries had not taken pains, in good earnest, to put such ideas out of their heads? It is notorious, that the savages believe that there are no extremities of barbarity, but what are within the rules of war against those whom they consider as their enemies. Inexpressible are the efforts which these same missionaries have employed to restrain, on such occasions, this criminal ferocity, especially as the savages deemed themselves authorized by right of reprisals. How many unfortunate persons of the English nation would have been detained for ever captives, or undergone the most cruel deaths, if, by the intervention of the missionaries, the savages had not been prevailed on to release them?

They are even ready to prove, by their written instructions, the lessons they inculcate to the savages, of the humanity and gentleness they ought to practise, even in time of war. It is especially ever since about seventeen years ago, that they do not cease declaiming against those barbarous and sanguinary methods of proceeding that seem innate to them. On this principle it is, that in the written maxims of conduct for them, care has been taken to insert a chapter, which, from the beginning to the end, places before their eyes the extreme horror they ought to have of such enormities. Their children particularly are sedulously taught this whole chapter, whence it comes, that one may daily perceive them growing more humane, and more disposed to listen, on this head, to the remonstrances of the missionaries.

[*To this plea of innocence in the French missionaries, as to any instigation of the savages to hostilities against the English, we shall oppose the testimony of their own court, in the following words of the French ministry, in the very same year,* 1751.

"His Majesty (the French king) has already observed, that the savages have hitherto been in the most *favorable dispositions*; and it even appears, that the conduct of the general C—n—ll—s, with respect to them, has only served to exasperate them more and more. It is of the *greatest importance*, both for the present and future, to keep them up to that spirit. The *missionaries* amongst them, are more than any one at hand to *contribute thereto*, and his majesty has *reason* to be *satisfied* with the *pains* they take in it. Our governor must excite these *missionaries* not to *slacken their endeavours* on this head. But he should advise them to *contain* their *zeal* within due bounds, so as not to render themselves *obnoxious* to the English, unless for very good purpose, and so as to avoid giving handle for just complaints.»

In this his most Christian Majesty has been faithfully served by these missionaries, in all points, except that political injunction of not giving a handle for just complaints, which they overshot in the ardor of their zeal; since it is undoubted matter of fact, that the missionaries openly employed all their arts, and all the influence of religion, to invenom the savages against us. Thence, besides a number of horrid cruelties, the most treacherous and base murder of captain How, at a conference, by some savages they set on, who perpetrated it within sight of the French forces. The publishing, however, of the foregoing memorial may have this good effect, that it will apprise the English of the matter of accusation against them, and enable them to counter-work those holy engines of state, and emissaries of ambition. It is also certain, that this very memorial was drawn up by a French priest, purely to furnish the French ministry a specious document to oppose to the most just representations of the British government. Besides the fictions with which it abounds, he has taken care to suppress the acts of cruelty committed, and the atrocious provocations given by the savages, at the instigation of his fellow-laborers sedition and calumny.]

LETTER FROM Mons. DE LA VARENNE, TO HIS FRIEND at ROCHELLE

Louisbourg **, the 8th of** *May*, **1756.**

Though I had, in my last, exhausted all that was needful to say on our private business, I could not see this ship preparing for France, especially with our friend *Moreau* on board, without giving you this further mark of how ardently I wish the continuance of our correspondence. It will also serve to supplement any former deficiencies of satisfaction to certain points of curiosity you have stated to me; this will give to my letter a length beyond the ordinary limits of one: and I have before-hand to excuse to you, the loose desultory way in which you will find I write, as things present themselves to my mind, without such method or arrangement, as a formal design of treating the subject would exact. But who looks for that in a letter?

I need not tell you how severely our government has felt the dismemberment of that important tract of country already in the possession of the English, under the name of Acadia; to say nothing of their further pretentions, which would form such terrible encroachments on Canada. And no wonder it should feel it, considering the extent of so fruitful, and valuable a country as constitutes that peninsula. It might of itself form a very considerable and compact body of dominion, being, as you know, almost everywhere surrounded by the sea, and abounding with admirable and well-situated ports. It is near one hundred leagues in length, and about sixty in breadth. Judge what advantages such an area of country, well-peopled, and well-cultivated, and abounding in mines, might produce. It is full of hills, though I could not observe any of an extraordinary heighth, except that of Cape Doree, at the mouth of the river *des Mines*, the most fertile part of it in corn and grain, and once the best peopled. There are a number of rivers very rapid, but not large, except that of St. John's, which is the finest river of all Acadia, where good water is rather scarce.

The soil in the vallies is rich, and even in the uplands, commonly speaking, good. The grains it yields are wheat, pease, barley, oats, rye, and Indian corn, and especially that of the vallies, for the higher ground is not yet cultivated. The pastures are excellent and very common, and more than sufficient to supply Cape-Breton, with the cattle that may be raised. There is fine hunting, and a plentiful fishing for cod, salmon, and other fish, particularly on the east-side, which is full of fine harbours at the distance of one, two, three, four, or of six or seven leagues at farthest from one another, within the extent of ninety leagues of coast. It is thought, in short, this fishery is better than any on the coasts belonging to France.

The air is extreamly wholesome, which is proved by the longevity of its inhabitants. I myself know some of above an hundred years of age, descendants from the French established in Acadia. Distempers are very rare. I fancy the climate is pretty near the same as in the north of China, or Chinese-Tartary. This country too, being rather to the southward of Canada, is not so cold as that; the snow not falling till towards St. Andrew's day: nor does it lie on the ground above two or three days at most, after which it begins to soften; and though the thaw does not take place, the weather turns mild enough to allow of working, and undertaking journeys. In short, what may be absolutely called cold weather, may be reduced to about twenty-five or thirty days in a winter, and ceases entirely towards the end of March, or at latest, the middle of April. Then comes the seed-time. Then are made the sugar and syrups of maple, procured from the juice or sap of that tree, by means of incisions in the bark; which sap is carefully received in proper vessels.

I could never find any ginseng-root; yet I have reason to believe there may be some in or near the hills, as the climate and situation have so much affinity to the northern provinces of China, or Northwest Tartary, as described to us by our missionaries.

We have very little knowledge of the medicinal herbs in this country, though some of them have certainly great virtue. There are the maiden-hair, the saxi-frage, and the sarsaparilla. There is also a particular root in this country of an herb called *Jean Hebert*, about the ordinary size of the *Salsifix*, or *Goatsbread*, with knots at about an inch, or an inch and an half distance from one another, of a yellowish colour, white in the inside, with a sugarish juice, which is excellent for the stomach.

There has been lately discovered in these parts a poisonous root, much resembling, in color and substance, a common carrot. When broke it has a pleasing smell; but between the flakes may be observed a yellowish juice, which is supposed to be the poison. Of four soldiers that had eaten of it in their soup lately, two were difficultly preserved by dint of antidotes; the other two died in the utmost agonies of pain, and convulsions of frenzy. One of them was found in the woods sticking by the head in a softish ground, into which he had driven it, probably in the excess of his torture. Such a vegetable must afford matter of curious examination to a naturalist; for as it does so much harm, it may also be capable of great good, if sought into by proper experiments.

The spirit of turpentine is much used by the inhabitants. The gum itself is esteemed a great vulnerary; and purges moderately those who are full of bilious, or gross humors.

For the rest there is, I believe, hardly any sort of grain, tree, or vegetable, especially in the north of France that might not be successfully raised in Acadia. The rains are frequent in every season of the year. There are indeed often violent squalls of wind, especially from the South, and seem the West, but nothing like the hurricanes in the West-Indies. It is a great rarity if thunder does any mischief. Some years ago there was a man killed in his hut by it; but the oldest men of the country never remembered to have known or heard of any thing like it before. There have been earthquakes felt but rarely, and not very violent. This country produces no venomous beasts, at least, that I could hear of. In the warmer season there are sometimes found snakes, not, however, thicker than one's finger, but their bite is not known to be attended with any fatal consequences, There are no tygers, nor lions, nor other beasts of prey to be afraid of unless bears, and that only in their rutting-time, and even then it is very rare that they attack. As there are then no carnivorous animals except the lynxes, who have a beautiful skin, and these rarely fall upon any living creatures; the sheep, oxen, and cows, are turned out into the woods or commons, without any fear for them. Partridges are very common, and are large-sized, with flesh very white. The hares are scarce, and have a white fur. There are a great many beavers, elks, cariboux, (moose-deer) and other beasts of the cold northern countries.

The original inhabitants of this country are the savages, who may be divided into three nations, the *Mickmakis*, the *Maricheets*, or *Abenaquis*, (being scarcely different nations) and the *Canibats*.

The *Mickmakis* are the most numerous, but not accounted so good warriors as the others: but they are all much addicted to hunting, and to venery; in which last, however, they observe great privacy. They are fond of strong liquors, and especially of brandy: that is their greatest vice. They are also very uncurious of paying the debts they contract, not from natural dishonesty, but from their having no notion of property, or of meum or tuum. They will sooner part with all they have, in the shape of a gift, than with any thing in that of payment. Honors and goods being all in common amongst them, all the numerous vices, which are founded upon those two motives, are not to be found in them. Yet it is true, that they have chiefs to whom they give the title of *Sagamo*; but all of them almost, at some time or other, assume to themselves this quality, which is never granted by universal consent, but to the personal consideration of distinguished merit in councils, or in arms. Their troops have this particularity, that they are, for the most part, composed of nothing but officers; insomuch that it is rare to find a savage in the service that will own himself a private man. This want of subordination does not, however, hinder them from concurring together in action, when their native ferocity and emulation stand them, in some sort, instead of discipline.

They are extreamly vindictive, of which I shall give you one example. Mons. *Daunay*, a French captain, with a servant, being overset in a canoe, within sight of some savages, they threw themselves into the water to save them, and the servant was actually saved. But the savage, who had pitched upon Mons. *Daunay*, seeing who it was, and remembering some blows with a cane he had a few days before received from him, took care to souse him so often in the water, that he drowned him before he got ashore.

It is remarked, that in proportion as the Europeans have settled in this country, the number of the savages considerably diminishes. As they live chiefly upon their hunting, the woods that are destroyed to cultivate the country, must in course contract the district of their chace, and cause a famine amongst them, that must be fatal to them, or compel them to retire to other countries. The English, sensible of this effect, and who seemed to place their policy in exterminating these savage nations, have set fire to the woods, and burnt a considerable extent of them. I have myself crossed above thirty leagues together, in which space the forests were so totally consumed by fire, that one could hardly at night find a spot wooded enough to afford wherewithal to make an extempore cabbin, which, in this country, is commonly made in the following manner: Towards night the travellers commonly pitch upon a spot as near a rivulet or river as they can; and as no one forgets to carry his hatchet with him, any more than a Spanish don his toledo, some cut down wood for firing for the night; others branches of trees, which are stuck in the ground with the crotch uppermost, over which a thatching is laid of fir-boughs, with a fence of the same on the weather-side only. The rest is all open, and serves for door and window. A great fire is then lighted, and then every body's lodged. They sup on the ground, or upon some leaved branches, when the season admits of it; and afterwards the table serves for a bed. The savages themselves rarely have any fixed hut, or village, that maybe called a permanent residence. If there are any parts they most frequently inhabit, it is only those which abound most in game, or near some fishing-place. Such were formerly for them, before the English had driven them away, *Artigoneesch*, *Beaubassin*, *Chipoody*, *Chipnakady*,

Yoodayck, Mirtigueesh, La Héve Cape Sable, Mirameeky, Fistigoisch, La Baye des Chaleurs Pentagony, Medochtek, Hokepack, and *Kihibeki.*

At present these savage nations bear an inveterate antipathy to the English, who might have easily prevented or cured it, if instead of rigorous measures, they had at first used conciliative ones: but this it seems they thought beneath them. This it is, that has given our missionaries such a fair field for keeping them fixed to the French party, by the assistance of the difference of religion, of which they do not fail to make the most. But lest you may imagine I am giving you only my own conjectures, take the following extract from, a letter of father Noel de Joinville, of a pretty antient date.

"I have remarked in this country so great an aversion in the convert-savages to the English, caused by difference of religion, that these scarce dare inhabit any part of Acadia but what is under their own guns. These savages are so zealous for the Roman Catholick church, that they always look with horror upon, and consider as enemies those who are not within the pale of it. This may serve to prove, that if there had been *priests* provided in time, to work at the conversion of the savages of New-England, before the English had penetrated into the interior of the county as far as they have done, it would not have been possible for them to appropriate to themselves such an extent of country as, at this day, makes of New-England alone the most magnificent colony on the face of the earth." [This pompous epithet might have yet been more just, if the improvement of that colony had been enough the care of the state, to have been pushed all the lengths of which it was so susceptible. Few Englishmen will, probably, on reflexion deny, that if but a third of those sums ingulphed by the ungrateful or slippery powers on the continent, upon interests certainly more foreign to England than those of her own colonies, or lavished in a yet more destructive way, that of corrupting its subjects in elections: if the third, I say, of those immense sums, had been applied to the benefit of the plantations, to the fortifying, encouraging, and extending them, there would, by this time, have hardly been a Frenchman's name to be heard of in North-America especially.]

But with this good father's leave, he attributes more influence to religion, though as the priests manage it, it certainly has a very considerable one, than in fact belongs to it. Were it not for other concurring circumstances that indispose the savages against the English, religion alone would not operate, at least so violently, that effect. Every one knows, that the savages are at best but slightly tinctured with it, and have little or no attachment to it, but as they find their advantage in the benefits of presents and protection, it procures to them from the French government. In short, it is chiefly to the conduct of this English themselves, we are beholden for this favorable aid of the savages. If the English at first, instead of seeking to exterminate or oppress them by dint of power, the sense of which drove them for refuge into our party, had behaved with more tenderness to them, and conciliated their affection by humoring them properly, and distributing a few presents, they might easily have made useful and valuable subjects of them. Whereas, disgusted with their haughtiness, and scared at the menaces and arbitrary encroachments of the English, they are now their most virulent and scarce reconcileable enemies. This is even true of more parts in America, where, though the English have liberally given presents to ten times the value of what our government does, they have not however had the same effect. The reason of which is clear: they make them with so ill a grace, and generally time their presents so unjudiciously, as scarce ever to distribute them, but just when they want to carry some temporary point with the savages, such, especially, as the taking up the hatchet against the French. This does not escape the natural sagacity of the savages, who are sensible of the design lurking at bottom of this liberality, and give them the less thanks for it. They do not easily forget the length of time they had been neglected, slighted, or unapplied to, unless by their itinerant traders, who cheat them in their dealings, or poison them with execrable spirits, under the names of brandy and rum. Whereas, on the contrary, the French are assiduously caressing and courting them. Their missionaries are dispersed up and down their several cantonments, where they exercise every talent of insinuation, study their manners, nature, and weaknesses, to which they flexibly accommodate themselves, and carry their points by these arts. But what has, at least, an equal share in attaching the savages to our party, is the connivence, or rather encouragement the French government has given to the natives of France, to fall into the savage-way of life, to spread themselves through the savage nations, where they adopt their manners, range the woods with them, and become as keen hunters as themselves. This conformity endears our nation to them, being much better pleased with seeing us imitate them, than ready to imitate us, though some of them begin to fall into our notions, as to trafficking and bartering, and knowing the use of money, of which they were before totally ignorant. We employ besides a much more effectual method of uniting them to us, and that is, by the intermarriages of our people with the savage-women, which is a circumstance that draws the ties of alliance closer. The children produced by these are generally hardy, inured to the fatigues of the chace and war, and turn out very serviceab subjects in their way.

But what is most amazing is, that though the savage-life has all the appearance of being far from eligible, considering the fatigues, the exposure to all weathers, the dearth of those articles which custom has made a kind of necessaries of life to Europeans, and many other inconveniencies to be met with in their vagabond course; yet it has such charms for some of our native French, and even for some of them who have been delicately bred, that, when once they have betaken themselves to it young, there is hardly any reclaiming them from it, or inducing them to return to a more civilized life. They prefer roving in the woods, trusting to the chapter of accidents for their game which is their chief support, and lying all night in a little temporary hut, patched up of a few branches; to all the commodiousness they might find in towns, or habitations, amongst their own countrymen. By degrees they lose all relish for the European luxuries of life, and would not exchange for them the enjoyments of that liberty, and faculty of wandering about, for which, in the forests, they contract an invincible taste. A gun with powder and ball, of which they purchase a continuation of supplies with the skins of the beasts they kill, set them up. With these they mix amongst the savages, where they get as many women as they please: some of them are far from unhandsome, and fall into their way of life, with as much passion and attachment, as if they had never known any other.

Mons. *Delorme,* whom you possibly may have seen in Rochelle, where he had a small employ in the marine-department, brought over his son here, a very hopeful youth, who had even some tincture of polite education, and was not above thirteen years old, and partly from indulgence, partly from a view of making him useful to the government, by his learning, at that age, perfectly the savage language, he suffered him to go amongst the savages. The young *Delorme* would, indeed, sometimes return home just on a visit to his family; but always expressed such an impatience, or rather pining to get back again to them, that, though reluctantly, the father was obliged to yield to it. No representations in short, after some years, could ever prevail on him to renounce his connexions, and residence amongst the *Abenaquis,* where he is almost adored. He has learned to excel them all, even in their own points of competition. He out-does them all in their feats of activity, in running, leaping, climbing mountains, swimming, shooting with the bow and arrow, managing of canoes, snaring and killing birds and beasts, in patience of fatigue, and even of hunger; in short, in all they most value themselves upon, or to which they affix the idea of personal merit, the only merit that commands consideration amongst them. They are not yet polished enough to admire any other. By this means, however, he perfectly reigns amongst them, with a power the greater, for the submission to it not only being voluntary, but the effect of his acknowledged superiority, in those points that with them alone constitute it. His personal advantages likewise may not a little contribute thereto, being perfectly well-made, finely featured, with a great deal of natural wit, as well as courage. He dresses, whilst with the savages, exactly in their manner, ties his hair up like them, wears a tomby-awk, or hatchet, travels with *rackets*, (or Indian shoes) and, in short, represents to the life the character of a compleat savage-warrior. When he comes to *Quebec*, or *Louisbourg*, he resumes his European dress, without the least mark appearing in his behaviour, of that wildness or rudeness one would naturally suppose him to have contracted by so long a habit of them with the savages. Nobody speaks purer French, or acquits himself better in conversation. He takes up or lays down the savage character with equal grace and ease. His friends have, at length, given over teazing him to come and reside for good amongst them; they find it is to so little purpose. The priests indeed complain bitterly, that he is not overloaded with religion, from his entering so thoroughly into the spirit of the savage-life; and his setting an example, by no means edifying, of a licentious commerce with their women; besides, his giving no signs of his over-respecting either their doctrine or spiritual authority. This they pretend hurts them with their actual converts, as well as with those they labor to make; though, in this conduct, he is not singular, for the French wood-rangers, in general, follow the like course in a greater or lesser degree. These representations of the priests would, however, have greater influence with our government, if the temporal advantage they derive from these rovers, undisciplined as they are, did not oblige them to wink at their relaxation in spirituals.

But it is not only men that have taken this passion for a savage life; there have been, though much rarer, examples of our women going into it. It is not many years since a very pretty French girl ran away into the woods with a handsome young savage, who married her after his country fashion. Her friends found out the village, or rather ambulatory tribe into which she had got; but no persuasions, or instances, could prevail on her to return and leave her savage, nor on him to content to it; so that the government not caring to employ force, for fear of disobliging the nation of them, even acquiesced in her continuance amongst them, where she remains to this day, but worshipped like a little divinity, or, at least, as a being superior to the rest of their women. Possibly too she is not, in fact, so unhappy, as her choice would make one think she must be; and if opinion constitutes happiness, she certainly is not so.

There are not wanting here, who defend this strange attachment of some of their countrymen to this savage life, on principles independent of the reason of state, for encouraging its subjects to spread and gain footing amongst the savage nations, by resorting to their country, of which they, at the same time, gain a knowledge useful to future enterprizes, by a winning conformity to their actions, and by intermarriages with them. They pretend, that even this savage life itself is not without its peculiar sweets and pleasures; that it is the most adapted, and the most natural to man. Liberty, they say, is no where more perfectly enjoyed, than where no subordination is known, but what is recommended by natural reason, the veneration of old age, or the respect of personal merit.

The chace is at once their chief employment and diversion; it furnishes them with means to procure those articles, which enter into the small number of natural wants. The demands of luxury, they think too dearly bought with the loss of that liberty and independence they find in the woods. They despise the magnificence of courts and palaces, in comparison with the free range and scope of the hills and vales, with the starry sky for their canopy: they say, we enjoy the Universe only in miniature, whilst the savage-rovers enjoy it in the great. Thus reason some of our admirers here of the savage-system of life, and yet I do not find that these refining advocates for it, are themselves tempted to embrace it. They are content to commend what themselves do not care to practise. Those who actually do embrace it, reason very little about it, though no doubt, the motives above assigned for their preference, are generally, one may say instinctively, at the bottom of it. Their greatest want is of wine, especially at first to those who are used to it; but they are soon weaned from it by the example of others, and content themselves with the substitution of rum, or brandy, of which they obtain supplies by their barter of skins and furs. In short, their hunting procures them all that they want or desire, and their liberty or independence supplies to them the place of those luxuries of life, that are not well to be had without the sacrifice in some sort of it.

It is more difficult to find an excuse for the shocking cruelties and barbarities, exercised by the savages on their unhappy captives in war. The instances, however, of their inhumanity, are certainly not exagerated, nor possible to be exagerated, but they are multiplied beyond the limits of truth. That they put then their prisoners to death by exquisite tortures, is strictly true; but it is as true too, that they do not serve so many in that manner as has been said. Numbers they save, and even incorporate with their own nation, who become as free as, and on a footing with, the conquerors themselves. And even in that cruelty of theirs, there is at the bottom a mixture of piety with their vindictiveness. They imagine themselves bound to revenge the deaths of their ancestors, their parents, or relations, fallen in war, upon their enemies, especially of that nation by whom they have fallen. It is in that apprehension too, they extend their barbarity to young children, and to women: to the first, because they fear they may grow up to an age, when they will be sure to pursue that revenge of which the spirit is early instilled into them; to the second, lest they should produce children, to whom they would, from the same spirit, be sure to inculcate it. Thus, in a round natural enough, their fear begets their cruelty, and their cruelty their fear, and so on, *ad infinitum*. They consider too these tortures as matter of glory to them in the constancy with which they are taught to suffer them; they familiarize to themselves the idea of them, in a manner that redoubles their natural courage and ferocity, and especially inspires them to fight desperately in battle, so as to prefer death to a captivity, of which the consequences are, and may be, so much more cruel to them. Another reason is also assignable for their carrying things to these extremities: War is considered by these people as something very sacred, and not lightly to be undertaken; but when once so, to be pushed with the utmost rigor by way of terror, joining its aid towards the putting the speediest end to it. The savage nations imagine such examples necessary for deterring one another from coming to ruptures, or invading one another upon slight motives, especially as their habitations or villages used to be so slightly fortified, that they might easily be surprised. They have lately indeed learned to make stronger inclosures, or pallisadoes, but still not sufficient entirely to invalidate this argument for their guarding against sudden hostilities, by the idea of the most cruel revenge they annex to the commission of them. It is not then, till after the maturest deliberation, and the deepest debates, that they commonly come to a resolution of *taking up the hatchet*, as they call declaring of war; after which, there are no excesses to which their rage and ferocity do not incite them. Even their feasting upon the dead bodies of their enemies, after putting them to death with the most excruciating tortures they can devise, is rather a point of revenge, than of relish for such a banquet.

That midst all their savageness they have, however, some glimmering perception of the *laws of nations*, is evident from the use to which they put the *calumet*, the rights of which are kept inviolate, thro' especially the whole northern continent of America. It answers nearest the idea of the olive-branch amongst the ancients.

As to your question, Sir, about the English being in the right or wrong, in their treatment of the *Acadians*, or descendants of the Europeans first settled in Acadia, and in their scheme of dispersing them, the point is so nice, that I own I dare not pronounce either way: but I will candidly state to you certain facts and circumstances, which may enable yourself to form a tolerably clear idea thereon.

But previously I shall give you a succinct description of these people: They were a mixed breed, that is to say, most of them proceeded from marriages, or concubinage of the savage women with the first settlers, who were of various nations, but chiefly French, the others were English, Scotch, Swiss, Dutch, &c. the Protestants amongst whom, and especially their children were, in process of time, brought over to a conformity of faith with ours. They found they could not easily keep their footing in the country, or live sociably with the great majority of the French, but by this means of coming over to our religion.

Certain Normans, of which number was Champlein, were the *first* French that discovered Port-Royal, now Annapolis, where they found some Scotch settled, who had built a fort of turf, and planted in the area before it some plumb-trees, and walnut-trees, which was all the works of agriculture, and fortification the British nation had made in this country before the year 1710. This is the chief reason [And a very good one surely.] too, why they so much insist on calling Acadia, Nova-Scotia, and pretend to be the first inhabitants and true proprietors. These Scotch were driven from Port-Royal by the Normans. It is true, they had discovered the river of Port-Royal *before* the Normans, and had built a turf-fort; but it is by no means true, that they were therefore the true settlers on this river, and less yet in the whole of Acadia. [Nothing can be more false and pitiful, than what follows of this Frenchman's reasoning. If a fort is not a settlement, what can be called one? Is it not one of the most valid, and generally received marks of taking possession? It supposes always a design to cultivate and improve; and no doubt but these first settlers would have done both, if they had not been untimely driven away.] The true inhabitants are those who cultivate a country, and thereby acquire a real permanent situation. The property of ground is to them who clear, plant, and improve it. The English had done nothing in this way to it till the year 1710. They never came there, but on schemes of incursion or trade; and in all the wars they had with the French, on being superior to them, they contented themselves with putting them to ransom; and though they sometimes took their fortified places, they did not settle in them. As all their pretension in Acadia was trade, they sometimes indeed detained such French as they could take prisoners; but that was only for the greater security of their traffic in the mean while with the savages. Traders, continually obliged to follow the savages in their vagabond journeys, could not be supposed to have time or inclination for agriculture. This title then the French settlers had; and in short, the whole body of the inhabitants of Acadia, from time immemorial, may be averred to have been French, since a few families of English, and other Europeans, cannot be said to form an exception, and those, as I have before observed, soon became frenchified. Except a few families from Boston or New-England I could never learn there were above three of purely British subjects, who also, ultimately conforming both in the religious and civil institutions to the French, became incorporated with them. These families were the *Peterses*, the *Grangers*, the *Cartys*. These last indeed descended from one Roger John-Baptist Carty, an Irish Roman-Catholic. He had been an indented servant in New-England, and had obtained at length his discharge from his master, with permission to remain with the French Acadians for the freer exercise of his religion. Peters was an iron-smith in England, and together with Granger, married in Acadia, and was there naturalized a Frenchman. Granger made his abjuration before M. Petit, secular-priest of the seminary of Paris, then missionary at Port-Royal (Annapolis). These and other European families then soon became united with the French Acadians, and were no longer distinguished from them. Most of these last were originally from *Rochelle*, *Xaintonge*, and *Poitou*; but all went under the common name of Acadians; and were once very numerous. The Parish of *Annapolis-Royal* alone in 1754, according to the account of father *Daudin*, contained three hundred habitations, or about two thousand communicants. The *Mines*, which are about five-and-thirty leagues from Port-Royal, and the best corn country in Acadia, were also very populous; nor were there wanting inhabitants in many commodious parts of this peninsula.

The character of the French Acadians was good at the bottom: their morals far from vitious; their constitution hardy, and yet strongly turned to indolence and inaction, not caring for work, unless a point of present necessity pressed them; much attached to the customs of the country, which have not a little of the savage in them, and to the opinions of their fore-fathers, which they cherished as a kind of patrimony; it was hard to inculcate any novelty to them. They had many parts of character in common with the Canada French. A little matter surprises, and sets them a staring, without stirring their curiosity to examine, or exciting their inclination to adopt or embrace it. They are remarkably fond of rosaries, crucifixes, agnus deis, and all the little trinkets consecrated by religion, with which they love to adorn their persons, and of which the priests make no little advantage in disposing of amongst them: and in truth, it is almost incredible what a power and influence these have over them, and with which they despotically govern them. One instance I am sure cannot but make you laugh. In September, 1754, the priest at *Pigigeesh*, had appointed his parishioners to perform the religious ceremony of a *Recess*, and to make them expiate some disgust they had given him, obliged them, men, women, and children, to attend the adoration of the holy-sacrament with a rope about their necks; and what is more, he not only made them all buy the rope of him, in which you may be sure he took care to find his account, but exacted their coming to fetch it bare-footed, from his parsonage house; and this they quietly submitted to. In short, considering the sweets of power on whomsoever exercised, our good fathers the missionaries are not so much to be pitied, as they would have us believe, for their great apostolical labors, and exposure to fatigue; since it is certain, they live like little kings in their respective parishes, and enjoy in all senses the best the land affords; and even our government itself, for its own ends, is obliged to pay a sort of court to them, and to keep them in good humour.

The Acadian men were commonly drest in a sort of coarse black stuff made in the country; and many of the poorer sort go bare-footed in all weathers. The women are covered with a cloak, and all their head-dress is generally a handkerchief, which would serve for a veil too, in the manner they tied it, if it descended low enough.

Their dwellings were almost all built in an uniform manner; the inhabitants themselves it was who built them, each for himself, there being but few or no mechanics in the country. The hatchet was their capital and universal instrument. They had saw-mills for their timber, and with a plane and a knife, an Acadian would build his house and his barn, and even make all his wooden domestic furniture. Happy nation! that could thus be sufficient to itself, which would always be the case, were the luxury and the vanity of other nations to remain unenvied.

Such in short were the French Acadians, who fell under the dominion of the king of Great Britain, when the English experienced, from both the Acadians and savages, a most thorough reluctance to the recognition of their new sovereign, which has continued to this day.

As to the savages it is certain, that the governors for the English acted entirely against the interest of their nation, in their procedure with them. They had been long under the French government, so far as their nature allows them to be under any government at all; and besides almost all the Micmakis, and great numbers of the Maricheets, or Abenaquis, were converted to our faith, and were consequently under the influence of the priests. It could not then be expected, naturally speaking, that these people could all of a sudden shake off their attachment to, and connexions with our nation; so that, even after the cession of Acadia, they continued, with a savage sulleness, to give marks of their preference of our government. This could not fail of giving the English umbrage; and their impatience not brooking either delays, or soothing them into a temper and opinion more favorable to them: they let it very early be seen, and penetrated by the savages, that they intended to clear the country of them. Nor would this exterminating plan, however not over-humane, have been perhaps wholly an impolitical one, if they had not had the French for neighbors, who, ever watchful and alert in concerning themselves with what past in those parts, took care underhand, by their priests and emissaries, to inflame them, and to offer them not only the kindest refuge, but to provide them with all necessaries of life, sure of being doubly repaid by the service they would do them, if but in the mischief they would do the English, to whom it was a great point with our government to make Acadia as uncomfortable, and as untenable as possible. It was no wonder then, that the savages, ill-used by the English, and still dreading worse from them, being constantly plied by our caresses, presents, and promises, should prefer our nation to that. I have before said, that religion has no great hold of these savages, but it could not be but of some weight in the scale, where their minds were already so exulcerated against those of a different one, whom they now considered as their capital enemies. You may be sure like-wise, our priests did not neglect making the most of this advantage, which the English themselves furnished them by their indiscreet management: for certain it is, that a few presents well placed, proper methods of conciliation, and a very little time, would have entirely detached the savages from our interest, and have turned the system of annoyance of the English against the French themselves. Some English governors indeed grew sensible of this, and applied themselves to retrieve matters by a gentler treatment, but the mischief was already done and irretrieveable; and our missionaries took care to widen the breach, and to keep up their spirit of hatred and revenge, by instilling into them the notions of jealousy, that such overtures of friendship, on the part of the English, were no better than so many snares laid to make them perish, by a false security, since they could not hope to do it by open violence. One instance may serve to show you the temper of these people: Some years ago the English officers being assembled at the *Mines*, in order to take a solemn recognition from them of the king of Great Britain, when a savage, a new convert, called *Simon*, in spite of all dissuasion, went himself alone to the English commander, and told him, that all his endeavours to get the king of England acknowledged, would be to no purpose; that, for his part, he should never pay any allegiance but to the king of France, and drawing a knife, said, "This indeed is all the arms I have, and with this weapon alone, I will stand by the king of France till death."

Yet, with all this obstinacy of sentiments, once more I dare aver, the savages would have been easily won over and attached to the English party, had these gone the right way about it: and I well know that the French, who knew best the nature of the savages, much dreaded it; and were not a little pleased to see the English take measures so contrary to their own interest, and play the game so effectually into our hands. In short, we took, as was natural, all the advantage of their indiscretion and over-sight.

I come now to the Acadians, or what may more properly be called the French Acadians. These would undoubtedly have proved very valuable subjects to the English, and extreamly useful to them in improving a dominion so susceptible of all manner of improvement as *Acadia,* (Nova-Scotia) if they could have been, prevailed on to break their former ties of allegiance to the king of France, and to have remained quietly under the new government to which they were now transferred. But from this they were constantly dissuaded, and withheld by the influence of our French priests, cantoned, amongst them [The letter-writer might have here added the infamous arts and falsities by which these emissaries of the French imposed on those bigotted deluded people, and to that end made religion a vile tool of state. They represented to these Acadians, that it was an inexpiable crime against their faith, to hold any commerce with heretics, and much more so to enter into their interests;—that there would be no pardon for them, either in the other world, or even in this, when the French should regain, as they certainly would, possession of a country ceded so much against the grain. In short, they succeeded but too well in keeping up the spirit of rebellion amongst those infatuated devotees of theirs, who remained sullen and refractory to all the advances the English made to gain them.], who kept them steady to our party. You may be sure our government did not fail of constantly inculcating the expediency of this conduct to our priests; who not only very punctually and successfully conformed to their instructions on this head, but very often in the heat of their zeal so much exceeded them, as to draw on themselves the animadversion of the English government. This answered a double end, of hindering that nation from finding those advantages in this country, by the prospect of which it had been tempted to settle in it, and of engaging it to consider Acadia itself, as something not material enough to think worth its keeping, at the expence which it must occasion, and consequently induce the English to be the readier to part with it again, on any future treaty of peace. This too is certain, that the French themselves knew neither the extent, nor the value of this country, till they were sensible of the improvements the English were projecting; and the use now so easy to discover might be made of so fine an establishment. But to return to the Acadians: It must be confest the English had, with respect to them, a difficult game to play. To force such a number of families, of which too such great use might have been made, to evacuate the country, seems at first both impolitic and inhuman. But then it must be considered, that these people were absolutely untractable as to the English, and thoroughly under the direction of priests in an interest quite opposite to theirs. To have taken those priests entirely from them, would have exasperated them yet more, and was, in fact, a measure repugnant to that spirit of toleration in religious matters, of which they boast, and to which it must be owned they constantly adhered, as to these people, both in speculation and practice.

[Might not this dilemma have been removed, by procuring for them priests, since priests they must have, from neutral nations, such as the Flemings, the Roman Swiss Cantons, &c. whom a very small matter of reward and encouragement would, it is probable, have fixed in the English interest? At least, they could not have the same motives for fomenting rebellious principles, as the French priests, who were set on by that government.]

None of the Acadians were ever molested purely for their religion; and even the priests of our nation were always civilly treated by them, whenever they had not reason to think they meddled in temporal matters, or stirred up their parishioners to rebellion. I have seen many of their own letters that acknowledge as much; so that upon the whole, I do not see that the English could do otherwise than they did, in expelling their bounds a people, who were constitutionally, and invincibly, a perpetual thorn in their side, whom they could at best look on as secret domestic enemies, who wanted nothing but an occasion to do them all the mischief in their power, and of whom, consequently, there could not, for their interest and safety, remain too few in the land.

In the mean time the French took special care to appear at least to receive with open arms those *refugees*, whom their fear or hatred of the English drove out of that country; they gave them temporary places of habitation, both for them and their cattle, besides provisions, arms, tools, &c. till they should fix a settlement in some part of the French dominions here, which they recommended especially in the island of, or on the banks of the river of St. John; but they were at first very loth to come to a determination. And surely, these unfortunate victims of their attachment to the French government deserved all the reparation in its power to give them, for what they had quitted for the sake of preserving allegiance to it, even after their country had been transferred to another sovereign. I cannot, however, consistently with truth say, they were received as kindly as they deserved, which probably bred that undetermination of their's to fix a new settlement, as they were pressed to do by the French government. They retained still a hankering after their old habitations: the temporary new ones were far from being equally agreeable or convenient; and even the ancient settlers in those places where these refugees were provisionally cantoned, began to make complaints of their encroaching upon them, and to represent their apprehensions of their becoming burthensome to them. Some of our people in power, more sollicitous for their own private interest, than for the public good, were but too remiss in relieving and comforting these poor people. This, at length, indisposed them so, that after very pathetic remonstrances on the hardship of their case, and the motives upon which they thus suffered, great numbers of them began to listen seriously to the proposals made them by the English, to return upon very inviting terms to the settlements they had quitted. In short, it required the utmost art of the missionaries, and even a kind of coercion from the military power, to keep them from accepting the English offers. For when they presented a petition to Mons. *de Vergor*, for leave to return to the English district, this commander, after having remonstrated to them that he could not grant their request, nor decide any thing of himself in a matter of that importance, was forced, at length, to declare to them, that he would *shoot* any man who should attempt to go over to the English. [It should here be remarked, that these very people had taken the oath of allegiance to the crown of England, agreeable to the tenor of the treaty of Utrecht. But the French, not content with harbouring these causeless malecontents, that were actually deserters over to them, kept continually, by means of the priests, plying such as staid behind with exhortations, promises, menaces, in short, with every art of seduction, to engage them to withdraw their sworn allegiance to their now lawful sovereign. In short, if all the transactions of the French in those parts were thrown into a history, it would lay open to the world such a scene of complicated villainy, rebellion, perjury, subornation of perjury, perfidiousness, and cruelty, as would for ever take from that nation the power of pluming itself, as it now so impudently does, on its sincerity, fairness, and moderation. The English, on the other hand, too conscious of the justice of their cause at bottom, have been too remiss in their confutation of the French falsities: content with being in the right, they cared too little for having the appearance of being so, as if the world was not governed by appearances.] Thus these poor people remained under this deplorable dilemma. Some of them too, had not even habitations to go back if they would: they had been forced into the measure of deserting their country, and passing over to the French side, by the violence of the Abbot de Loutre, who had not only preached them into this spirit, but ordered the savages, whom he had at his disposal, to set fire to their habitations, barns, &c. particularly at *Mirtigueesh*. [The reader is desired to observe, that in the memorials delivered into the English court by the French ministers, this burning of villages was specifically made an article of complaint, at the same time that it was their own incendiary agent, at their own instigation, who had actually caused fire to be set to them by his savages. Could then impudence be pushed farther than it was on this occasion?]

In the mean time the French did not spare, at least, the consolation of words and promises to these distrest Refugee-acadians. They were assured, that they would infallibly be relieved on the regulation of the limits taking place, which was then on the point of being settled, by commissaries, between the two crowns. [The truth is, that in these assurances the French government, which never intended a conclusion, but only an amusement, did not scruple equally deceiving the English, and these infatuated Acadian subjects of ours, who, to the French interest had sacrificed their own, their possessions in their country, their sworn faith, in short, their ALL. Whoever has the patience to go through the French memorials, in their procedure with our commissaries, may see such instances of their pitiful prevarications, petty-fogging chicanery, quirks, and evasions, as would nauseate one. The whole stress of their argument, in short, turns merely upon names, where the things themselves were absolutely out of the question, from the manifest notoriety of them.] This hope, in some sort, pacified them; and they lived as well as they could in the expectation of a final decision, which was not so soon to come.

Yet even this example of the sufferings of these people, purely on account of their attachment to the French government, could not out balance with the French Acadians, who remained in the English district, the assiduous applications of our priests to keep them firm in the French interest. They never ceased giving every mark in their power of their preference of our government to that, under which the treaty of Utrecht had put them. The English, however, at length finding that, neither by fair nor foul means, could they reclaim or win them over to their purpose, so as that they might in future depend upon them, came at once to a violent resolution. They surprized and seized every French Acadian-man they could lay their hands on, (the women they knew would follow of course) and, to clear the country effectually of them, dispersed them into the remotest parts of their other settlements in North-America, where they thought they could do the least mischief to them. Some were shipped off for England: the priests shared the same fate, and were conveyed to Europe. With this evacuation, the very existence of the French Acadians may be said to have ended; for in Acadia there are scarce any traces of them left, few or none having escaped this general seizure and transportation, for the necessity of which, the English were perhaps more to be pitied than blamed.

In the mean time our government had so far succeeded, as to force the English, thus to deprive themselves of such a number of subjects, who, but for the reasons above deduced, might have been very valuable ones, and a great strengthening of their new colony. Hitherto then our neighborhood has made it almost as irksome, and uncomfortable to them, as we could wish; and this fine spot of dominion does not nigh produce to them the advantages that might otherwise naturally be expected from it. Numbers of themselves begin to exclaim against it, as if its value and importance had been overrated; not considering, that it is on the circumstances of their possession, and not on the nature of the possession itself, that their complaints and murmurings should fall. It is very likely, that whenever we get it back again, we shall know very well what to do with it. They have begun to teach us the value of what we thus inadvertently parted with to them; and it will be hard, indeed, on recovering it, if we do not improve upon their lessons.

In the mean time you in Europe are cruelly mistaken, if you do not annex an idea of the highest consequence and value, to the matters of dominion now in dispute, between the crowns of France and Great Britain, between whom the war is in a manner begun, by the capture of the Alcides and Lys, and which, even without that circumstance, was inevitable. I know that our (French) government, is indeed fully sensible of the capital importance to it of its interest in these parts, and has proceeded in consequence. But it is not so, I find by your letters, and the reports of others, with numbers in Europe, who do not conceive, that the present object of the war is so considerable as it really is.

To say nothing of the vast extent of country that falls under the claim of the English to Acadia (Nova-Scotia) which alone would form an immence mass of dominion, greatly improveable in a number of points, its situation is yet of greater weight. By the English possessing it, Canada itself would be so streightened, so liable to harrassment, and especially to the comptrol of its navigation, that it would scarce be tenable, and surely not worth the expence of keeping. The country pretended to have been ceded is far preferable to it; and the masters of it would be equally masters of the sea all over North-America. Hallifax, for example, according to which of the nation's hand it should be in, may be equally an effectual check on Quebec, or Boston.

You will then allow, that was there even nothing more in dispute than the limits of the cession of Acadia, or Nova-Scotia, together with its necessary dependence, that alone would form such a considerable object, as not easily to be given up on either side. The commissaries appointed by both crowns, then failing of coming to any agreement or regulation, it is no wonder to see the appeal lodged with the sword; especially when there is another point yet remains, of perhaps equal, if not superior, importance, depending on the issue of the war: and that is, the western inland frontiers of the English colonies. Should we ever command the navigation of the lakes and rivers, behind their settlements, you can easily figure to yourself, not only the vast advantages of preserving that communication of Canada, with New Orleans and the Mississippi, so absolutely essential to both these our colonies, but the facility it will give us on all occasions of distressing the English, where neither their marine-force can succor them, nor can they be able to resist the attack, since we may make it wherever ever we please, and effectually dodge any land-force they might assemble in any one or two parts to oppose us. We may then carry the war into the quarter most convenient; and most safe for us, if we should ever have the whole navigation of the lakes so far at our disposal, as to prevent their constructing any material number vessels to dispute it with us. Thus we can penetrate into the heart of any of their colonies, that may best suit us, especially with the concurrent aid of the savages, whom we have found means to attach so strongly to us, and on whom we can greatly depend for the effectual harrassment of, especially, the back-plantations of the English.

You see then, Sir, by this summary sketch of the points in contest, that the war being once engaged, it will not be so easy a matter as many in Europe imagine, to adjust the pretensions, so various and so important, of the respective nations, so as to be able to procure a peace. Some, of the points appear to me absolutely *untreatable*. You may observe too, that I do not so much as touch upon the dispute about Tabago, Santa-Lucia, or any of the Leeward islands, which are not, however, of small consequence. In short, the war must, in all human probability, be a much longer one, than is commonly believed. Neither nation can materially relax of its claims, without such a thorough sacrifice of its interest in America, as nothing but the last extremities of weakness can compel.

Long as this letter is, I cannot yet close it without mentioning to you a singular phenomenon of nature, in the island of St. John. You know it is a flat, level island, chiefly formed out of the congestion of sand and soil from the sea. Tradition, experience, and authentic public acts (*Procés verbaux*) concur to attest that every seven years, it is visited by swarms either of locusts, or of field-mice, alternately, never together; without its being possible to discover hitherto either the reason, or the origin of these two species, which thus in their turns, at the end of every seventh year, pour out all of a sudden in amazing numbers, and having committed their ravages on all the fruits of the earth, precipitate themselves into the sea. Neither has any preventive remedy for this evil been yet discovered. It is well known how they perish, but, once more, how they are produced no one, that I could learn, has as yet been able to trace. The field-mice are undoubtedly something in the nature of those swarms of the sable-mice, that sometimes over-run Lapland and Norway, though I do not know that these return so regularly, and at such stated periods, as those of this island.

I am, Sir,

Your most obedient,

Humble servant.

CHARACTER OF THE SAVAGES of NORTH-AMERICA, EXTRACTED FROM A LETTER of the Father CHARLEVOIX, TO A LADY of Distinction

To give you, Madam, a summary sketch of the character of the savages in this country, I am to observe to you, that under a savage appearance, with manners and customs, that favor entirely of barbarism, may be found a society exempt from almost all the faults that so often vitiate the happiness of ours.

They appear to be without passion, but they are in cold blood, and sometimes even from principle, all that the most violent and most unbridled passion can inspire into those, who no longer listen to reason.

They seem to lead the most miserable of lives, and they are, perhaps, the only happy of the earth. At least those of them are still so, amongst whom the knowledge of those objects that disturb and seduce us, has not yet penetrated, or awakened in them, those pernicious desires which their ignorance kept happily dormant: it has not, however, hitherto made great ravages amongst them.

There may be perceived a mixture in them of the most ferocious and the most gentle manners; of the faults reproachable to the carnivorous beasts, with those virtues and qualities of the head and heart, that do the most honor to human-kind.

One would, at first, imagine, that they had no sort of form of government, that they knew no laws nor subordination, and that living in an entire independence, they suffered themselves to be entirely guided by chance, or by the most wild, untamed caprice: yet they enjoy almost all the advantages, which a well-regulated authority can procure to the most civilized nations. Born free and independent, they hold in horror the very shadow of despotic power; but they rarely swerve from certain principles and customs, founded upon good-sense, which stand them in the stead of laws, and supplement in some sort to their want of legal authority. All constraint mocks them; but reason alone hold them in a kind of subordination, which, for its being voluntary, does not the less answer the proposed end.

A man, whom they should greatly esteem, would find them tractable and ductile enough, and might very nearly make them do any thing he had a mind they should; but it is not easy to gain their esteem to such a point. They grant it only to merit, and that merit a very superior one, of which they are as good judges as those, who, amongst us, value themselves the most upon being so. They are, especially, apt to be taken with physiognomy; and there are not in the world, perhaps, men who are greater *connoisseurs* in it: and that is, because they have for no man whatever, any of those respects that prejudice or impose on us, and that studying only nature, they understand it well. As they are not slaves to ambition or interest, those two passions that have chiefly cancelled in us that sentiment of humanity, which the author of nature had engraved in our hearts; the inequality of conditions is not necessary to them, for the support of society.

There are not therefore, Madam, to be seen amongst them, or at least, are rarely to be met with, those arrogant haughty characters, who, full of themselves of their greatness, or their merit, look on themselves almost as a species a-part, and disdain the rest of mankind, of whom consequently they can never have the confidence or love. Their equals these rarely know any thing of, because the jealousy that reigns amongst the great, hinders them from being intimate enough with one another. Neither do they know themselves, from their never studying themselves, and from their constant self-flattery. They never reflect, that to gain admission into the hearts of men, they must make themselves their equals; so that with this pretended superiority of enlightened understanding, which they look on as an essential property of the rank they hold, the most part of them live groveling in a proud and incurable ignorance of all that it would be the most important for them to know, and never enjoy the true sweets of life.

In all this how wretchedly different from the savages! In this country, all the men esteem themselves equally men; and in man, what they most esteem is, the man. No distinction of birth; no prerogative attributed to rank, to the prejudice of the other free members of society; no pre-eminence annexed to merit that can inspire pride, or make others feel too much their inferiority. There is, perhaps, less delicacy in their sentiments than amongst us, but surely more uprightness; less ceremony; less of all that can form a dubious character; less of the temptations or illusions or self-love.

Religion only can perfect these people in what is good in them, and correct what bad. This indeed is not peculiar to them, but what is so, is, that they bring with them fewer obstacles to religious devotion when once they have begun to believe, which can only be the effect of a special grace. It is also true, that to establish firmly the empire of religion over them, it would be necessary that they should see it practised in all its purity by those who profess it. They are extremely susceptible of the scandal given by bad Christians, as are all those who are, for the first time, instructed in the principles of the Gospel-morality.

You will perhaps ask me, Madam, if they have a religion? To this I answer, that it cannot be said they have not one, though it is difficult to give a definition of what it is. I shall sometime or other, take occasion to enter into more particulars on this head. This letter, like most of the others that have preceded it, prove sufficiently that I do not pretend to write to you methodically.

I shall then now only content myself with adding, by way of finishing, to this picture of the savages, that even in their most indifferent actions, may be perceived the traces of the primitive natural religion, but which escape those who do not study them enough, because they are yet more defaced by the want of instruction, [This want of instruction is wretchedly supplemented amongst the savage-converts to the Popish religion, by that superstitious worship, and those fabulous traditions, its missionaries have introduced amongst them, and which must be only the more execrable, for their being a superstructure on so fair a foundation as that of the truths of the Gospel. At least, the savages, in their genuine unsophisticated state, have no such base, absurd, derogatory ideas of the Deity, as are implied by the doctrines of transubstantiation, purgatory, absolution, and the like fictions in the Romish church, which have been the more than mines of Mexico and Peru, of its clergy.] than adulterated by the mixture of a superstitious worship, and by fabulous traditions.